RESIDUE REVIEWS

RESIDUES OF PESTICIDES AND OTHER FOREIGN CHEMICALS IN FOODS AND FEEDS

RÜCKSTANDS-BERICHTE

RÜCKSTÄNDE VON PESTICIDEN UND ANDEREN FREMDSTOFFEN IN NAHRUNGS- UND FUTTERMITTELN

EDITED BY

FRANCIS A. GUNTHER

RIVERSIDE

VOLUME 12

SPRINGER-VERLAG
BERLIN · HEIDELBERG · NEW YORK
1966

ISBN 978-1-4615-8406-3 ISBN 978-1-4615-8404-9 (eBook)
DOI 10.1007/978-1-4615-8404-9

 Library of Congress Catalog Card Number 62—18595.
Softcover reprint of the hardcover 1st edition 1966

Title No. 6614

RESIDUE REVIEWS

RESIDUES OF PESTICIDES AND OTHER FOREIGN CHEMICALS IN FOODS AND FEEDS

RÜCKSTANDS-BERICHTE

RÜCKSTÄNDE VON PESTICIDEN UND ANDEREN FREMDSTOFFEN IN NAHRUNGS- UND FUTTERMITTELN

EDITED BY

FRANCIS A. GUNTHER

RIVERSIDE

VOLUME 12

SPRINGER-VERLAG · NEW YORK, INC.

1966

ISBN 978-1-4615-8406-3 ISBN 978-1-4615-8404-9 (eBook)
DOI 10.1007/978-1-4615-8404-9

Title No. 9755

In memoriam
Professor Dr. Maria Ester Alessandrini
Member Advisory Board, Residue Reviews,
Volumes 1—12

Preface

That residues of pesticide and other "foreign" chemicals in foodstuffs are of concern to everyone everywhere is amply attested by the reception accorded previous volumes of "Residue Reviews" and by the gratifying enthusiasm, sincerity, and efforts shown by all the individuals from whom manuscripts have been solicited. Despite much propaganda to the contrary, there can never be any serious question that pest-control chemicals and food-additive chemicals are essential to adequate food production, manufacture, marketing, and storage, yet without continuing surveillance and intelligent control some of those that persist in our foodstuffs could at times conceivably endanger the public health. Ensuring safety-in-use of these many chemicals is a dynamic challenge, for established ones are continually being displaced by newly developed ones more acceptable to food technologists, pharmacologists, toxicologists, and changing pest-control requirements in progressive food-producing economies.

These matters are also of genuine concern to increasing numbers of governmental agencies and legislative bodies around the world, for some of these chemicals have resulted in a few mishaps from improper use. Adequate safety-in-use evaluations of any of these chemicals persisting into our foodstuffs are not simple matters, and they incorporate the considered judgments of many individuals highly trained in a variety of complex biological, chemical, food technological, medical, pharmacological, and toxicological disciplines.

It is hoped that "Residue Reviews" will continue to serve as an integrating factor both in focusing attention upon those many residue matters requiring further attention and in collating for variously trained readers present knowledge in specific important areas of residue and related endeavors; no other single publication attempts to serve these broad purposes. The contents of this and previous volumes of "Residue Reviews" illustrate these objectives. Since manuscripts are published in the order in which they are received in final form, it may seem that some important aspects of residue analytical chemistry, biochemistry, human and animal medicine, legislation, pharmacology, physiology, regulation, and toxicology are being neglected; to the contrary, these apparent omissions are recognized, and some pertinent manuscripts are in preparation. However, the field is so large and the interests in it are so varied that the editor and the Advisory Board earnestly solicit suggestions of topics and authors to help make this international book-series even more useful and informative.

"Residue Reviews" attempts to provide concise, critical reviews of timely advances, philosophy, and significant areas of accomplished or needed endeavor in the total field of residues of these chemicals in foods, in feeds, and in transformed food products. These reviews are either general or specific, but properly they may lie in the domains of analytical chemistry and its methodology, biochemistry, human and animal medicine, legislation, pharmacology, physiology, regulation, and toxicology; certain affairs in the realm of food technology concerned specifically with pesticide and other food-additive problems are also appropriate subject matter. The justification for the preparation of any review for this book-series is that it deals with some aspect of the many real problems arising from the presence of residues of "foreign" chemicals in foodstuffs. Thus, manuscripts may encompass those matters, in any country, which are involved in allowing pesticide and other plant-protecting chemicals to be used safely in producing, storing, and shipping crops. Added plant or animal pest-control chemicals or their metabolites that may persist into meat and other edible animal products (milk and milk products, eggs, etc.) are also residues and are within this scope. The so-called food additives (substances deliberately added to foods for flavor, odor, appearance, etc., as well as those inadvertently added during manufacture, packaging, distribution, storage, etc.) are also considered suitable review material.

Manuscripts are normally contributed by invitation, and may be in English, French, or German. Preliminary communication with the editor is necessary before volunteered reviews are submitted in manuscript form.

Department of Entomology
University of California
Riverside, California
January 1, 1966

F. A. G.

Table of Contents

Correction to page 120
Residue Reviews, Volume 11

The graph which is now over the Figure 2 caption should appear over the Figure 1 caption and vice versa

The analytical determination of strontium-90 in foods

By
Sidney Davis*

Contents

I. Introduction

The interest in analytical methods for the determination of strontium-90 in foods can be attributed to the recognition of its presence in the food chain as a possible dietary hazard to man. Five criteria are generally cited to evaluate a radioactive contaminant as a possible dietary hazard. These are:

a) Its availability in significant quantity in soil. According to a *United Nations* report in 1957, the strontium-90 values in the soil of the United States varied from approximately 5.2 to 44.6 millicuries per square mile in the upper two inches of earth.

b) The efficiency of transfer through the food chain to the human diet.

c) The extent of assimilation by one or more specific body tissues.

d) Long biological half life.

e) Long physical half life.

It has been estimated that the fraction of foliar retention of strontium-90 ranges from 90 percent for pasture grass to much smaller values for rapidly growing, quickly harvested crops.

Radionuclides in water gain access to the diet through aquatic plants and animals, either by surface contamination or by biological uptake. Strontium-90 follows calcium approximately through the food chain from the environment and is eventually incorporated with calcium in the bone.

* Division of Pharmacology, Food and Drug Administration, U. S. Dept. of Health, Education, and Welfare, Washington 25, D. C.

During its residence for years in bone, it will continue to bombard the bone structure with its associated beta-ray emission. According to a *United Nations* report of 1957, it has been estimated that the concentration of strontium-90 per gram of calcium in the bone of man, due to milk consumption, would be about eight percent of that of soil.

II. The chemistry of strontium and yttrium applicable to strontium-90 determinations

The various analytical methods for strontium-90, or yttrium-90, the daughter of strontium-90, may be placed in three categories:

a) The classical series of gravimetric separations to remove impurities with a final precipitation of an insoluble salt of strontium or yttrium.

b) Ion exchange or other chromatographic methods.

c) Liquid-liquid partition techniques.

The chemistry of strontium, applicable to the conventional separation techniques, refers particularly to its position in group 2*A* of the periodic table as a member of the alkaline earth series, including the elements beryllium, magnesium, calcium, strontium, barium, and radium. Under certain conditions the insoluble alkaline earth compounds such as the carbonates, oxalates, chromates, sulfates, phosphates, and nitrates are used to separate the members of the group. Fortunately, the solubilities of some of these salts vary under certain specific conditions. With the selection of proper conditions of precipitation it is possible to separate members of this group or exclude impurities.

Some of the insoluble salts of strontium most used in strontium-90 determinations are the carbonates, oxalates, nitrates, and sulfates. The precipitation of strontium carbonate is non-specific in respect to the other members of the alkaline earths, but does result in a separation from the soluble alkali carbonates of cesium-137 and potassium-40. Precipitation is often used as a preliminary concentration step from large volumes or even as a final step for counting after chemical purification from other contaminants. For large volumes, adjustment to pH 10 or higher is made before adding sodium carbonate solution as a precipitating agent. The main advantage of strontium carbonate is its solubility in acid. The insoluble sulfate of strontium may be converted to strontium carbonate by heating with a solution of sodium carbonate. The resulting precipitate, when dissolved in acid, usually contains a minimum of impurities.

In many of the conventional gravimetric radiochemical separation techniques for strontium-90 determination an iron hydroxide scavenge is used to remove traces of a number of rare earths. Since strontium carbonate is insoluble in basic solution, precautions must be taken to insure that carbonate-free ammonium hydroxide is used as a precipitating agent for ferric hydroxide. Without this precaution strontium-90 may be lost in the ferric hydroxide scavenge.

Strontium oxalate is insoluble over a wide pH range. Strontium oxalate, monohydrate, and other alkaline earth oxalates may be precipitated at pH 2 to 3, leaving phosphates in solution. For samples of high phosphate content, the pH range of 2 to 3 for oxalate precipitation may be used as a preliminary separation. In procedures using ferric hydroxide scavenge of rare earths, the alkaline earth phosphates, being insoluble in basic solution, would be lost to the ferric hydroxide precipitate.

The insolubility of strontium nitrate in 60 to 80 percent nitric acid is used to separate strontium from the remainder of the alkaline earth elements. Most of the other fission products, including the actinides, are soluble in 70 percent nitric acid, offering a good gross decontamination step. The separation of calcium in foodstuffs is best with 60 percent nitric acid, but the precipitation of strontium is not complete. It is usual to perform the separation at a concentration of 70 percent nitric acid; then the re-precipitation is done several times to insure maximum decontamination from calcium.

Strontium sulfate is frequently used as the final precipitate before counting, even though it is not as insoluble as barium sulfate. Ethyl alcohol may be added to lower the solubility during the course of the precipitation. Strontium sulfate can be a source of loss of carrier in the original sample if the concentration of sulfate ion in the food ash is high. In these instances, strontium sulfate may be converted to strontium carbonate with sodium carbonate.

Selective precipitation of barium chromate occurs at pH 5.5. Radium co-precipitates with the barium nuclides under these conditions. However, strontium chromate does not start precipitating from solution until the solution becomes more basic.

The ferric hydroxide scavenge is useful for the removal of rare earth fission products, yttrium, and zirconium. Complete decontamination of strontium solutions by these nuclides is attained by a second ferric hydroxide precipitation or, as a substitute, a cerium-zirconium hydroxide scavenge. The second precipitation is a marking point for the beginninig of yttrium-90 ingrowth from strontium-90 samples.

The second precipitation is a marking point for the beginning of yttrium-90 ingrowth in food samples. The chemical properties of yttrium salts are quite similar to those of the rare earth lanthanides. It is covered completely in the *Nuclear Science Series* by Stevenson and Nervik (1961). Since the gravimetric separation techniques utilize ferric hydroxide scavenge to remove first ingrown yttrium-90 and the rare earth fission products, the second ingrown yttrium-90 from strontium-90 is free of rare earths.

Yttrium forms a number of organic complexes which are useful in liquid-liquid partition methods for the separation of yttrium-90. The method of Porter and Kahn (1964) makes use of solvent extraction from 14 N nitric acid into tributyl phosphate to separate yttrium from the lanthanum-140 daughter product of barium-140, after preliminary ion-exchange separations. More detailed discussion of the various solvent extraction methods follow later in this review.

From data presented by Hallden and Harley (1957), the ratio Sr-89/Sr-90, from one day to 300 days after atomic detonation, varies from 179 to 3.09. Therefore, the presence of strontium-89 in a food sample will complicate the strontium-90 assay. Strontium-89 and strontium-90 cannot be chemically separated, and the health hazards associated with the two isotopes differ greatly.

The solution to this problem lies in the decay schemes for both strontium isotopes:

$$\text{Sr-89} \xrightarrow[\text{50.5-day half life}]{\bar{\text{B}}} \text{Yttrium-89 (non-radioactive)}$$

$$\text{Sr-90} \xrightarrow[\text{27.7-year half life}]{\bar{\text{B}}} \text{Y-90} \xrightarrow[\text{64.2-hour half life}]{\bar{\text{B}}} \text{Zr-90 (non-radioactive).}$$

Since yttrium-89 is non-radioactive and yttrium-90, the daughter of strontium-90, is radioactive, the strontium-90 activity of a food sample is obtained by counting the yttrium-90 activity in a separated, purified, insoluble salt of yttrium. Usually, in the actual analysis, the ingrown yttrium is removed completely from the strontium sample. A new ingrowth period of 14 days is allowed to pass. During this period, radioactive secular equilibrium is attained to the extent of 97 percent. The newly ingrown yttrium-90 is separated and beta-counted. With suitable correction factors, the quantity of strontium-90 in the sample is calculated. When it is necessary to determine strontium-89, beta-counting of the total strontium fraction, usually the carbonate, is used. This precipitate contains strontiums-89 and 90 and ingrown yttrium-90. It is possible to calculate the strontium-89 content of the samples by the following set of calculations:

a) Determine the activity of Y-90 grown into the strontium fraction during the period between separation of Sr-90 to counting of sample. This activity, subtracted from the total activity, results in a value for total radiostrontium (Sr-89 + Sr-90).

b) The activity of Sr-90 in the sample is equal to the activity of Y-90 at secular equilibrium. This is obtained by applying suitable correction factors for the activity of Y-90 found in the yttrium fraction. This value is subtracted from the total radiostrontium value. The difference is the activity of Sr-89 in the sample.

The separation technique for the processing of liquid milk samples is more varied than that of more solid samples. Three methods of removal of organic material from milk have been reported. MURTHY *et al.* (1960) remove protein from milk by the addition of a solution of trichloroacetic acid. The lowering of the pH in this fashion results in a shift of protein-bound strontium to ionic strontium with an accompanying precipitation of the milk protein which is removed by filtration. The second method of separation involves ion-exchange techniques as described by PORTER *et al.* (1961 and 1964). In the earlier paper strontium and the other alkaline earth ions are absorbed by the cation-exchange resin as milk is passed through the column. The effluent material containing the organic and anionic material is discarded. The cations (Sr, Ca, Ba, Ra, Cs, and K) are eluted from the column, and chemical separations are used in order to separate the strontium-90. The later modification of this ion-exchange technique involves passing the milk sample consecutively through cation- and anion-exchange resin columns. The alkaline earth ions in milk are retained on the cation exchanger, while yttrium is retained on the anion exchanger as an anionic complex. The method of removal of organic material involves the conventional dry-ashing technique described in an *Atomic Energy Commission* publication (NYO-4700, August 1959). The organic material in milk is destroyed by ashing the sample, after drying, at 450 to 600 °C. The ash is solubilized with either acid leaching or fusion with sodium hydroxide—sodium carbonate. This method is time consuming and is the least preferred method.

The processing of fruits, vegetables, and meats is described in the publication mentioned above and three 1964 publications of the Radiological Health Laboratories of the *Public Health Service*. The technique involves cutting the samples into small portions, blending, and drying them in an oven at 105 to 110 °C. for approximately 12 hours. The dried samples, after quantitative transfer to evaporating dishes, are ashed in a muffle furnace at 450 to 600 °C. As with milk ash, the food ash is solubilized either with acid or fusion with sodium hydroxide — sodium carbonate.

Solvent extraction procedure have become most popular for the separation of radionuclides. Of the many organic chelating agents, thenoyl trifluoracetone, tributyl phosphate, and di-(2-ethyl-hexyl) phosphoric acid have been most generally used for the separation of yttrium-90.

III. Gravimetric methods of separation

The Harley (1963) method of radiochemical analysis for strontium-90 has been described for water, milk, soil, urine, bone, tissue, and vegetation. Strontium is separated from calcium, sodium, potassium, cesium, and rubidium by fuming nitric acid separation. Radium, lead, and barium are removed by barium chromate scavenging. Other fission products, including the lanthanides and ingrown yttrium-90, are removed with yttrium hydroxide precipitation. After the strontium-90, yttrium-90 secular equilibrium is reached in 14 days, the yttrium-90 is precipitated as the hydroxide, and is converted to the oxalate prior to the counting procedure. Specific directions for ashing various types of environmental material are given in detail.

From the standpoint of routine operation, the repeated use of fuming nitric acid separation can be a most disagreeable one for the analyst. Another disadvantage is encountered in samples of low activity containing large amounts of calcium. Repeated nitric acide separations may result in low strontium recoveries reflecting very low strontium-90 and yttrium-90 activities. In routine operations losses may be so high that the samples being counted may have activities of the same order of magnitude as background activity.

Welford and Sutton (1960) have devised a rapid screening technique for the estimation of total radiostrontium in milk. This attempt to break away from the conventional nitric acid separation of calcium from strontium makes use of the chelating properties of ethylenediaminetetraacetic acid (EDTA) with the alkaline earth cations. In weakly acidic solution,

EDTA does not react with the heavier alkaline earths. The raw milk is digested under pressure, and the mixture is filtered. Strontium and other insoluble alkaline earth phosphates are precipitated and collected from the filtrate, chelated with the tetra sodium salt of EDTA, and separated by alteration of pH. At pH 6.5, barium and radium are ionized and may be precipitated by addition of chromate ion. Strontium is ionized at pH 5 and may be precipitated as the sulfate. Calcium sulfate is relatively soluble in an environment of ammonium ions. Therefore, calcium discrimination is enhanced by introduction of chromate and sulfate ions as ammonium salts. The techniques were developed from prior published reports of FARABEE (1958) and AFANASIEVA (1959). Strontium yields were reported as 70 to 80 percent from the sulfate with less than one percent contamination from calcium. The estimate of strontium-90 content is a maximum value, since strontium-89 is also included in the final activity of the sample.

The contamination of the final counted precipitate by yttrium-91, promethium-147, cerium-144, zirconium-95, cesium-137, and ruthenium-106 ranges from approximately 0.1 to 0.001 percent. The method appears to be suitable as an emergency screening method.

The method of the *World Health Organization* (1959) follows a familiar pattern resembling the HARLEY (1963) technique. Vegetable ash is treated with dilute nitric acid in the presence of strontium carrier. The resulting filtered solution is nearly evaporated to dryness along with perchloric acid. Silica is filtered off. Calcium and strontium are precipitated as phosphate. Strontium is separated from calcium by a number of nitric acid precipitations. Barium is removed by chromate scavenge followed by ferric hydroxide scavenge to remove radium and daughter products. Yttrium is ingrown completely into the strontium fraction. Yttrium is separated as the hydroxide and oxalate while strontium is counted as the carbonate.

A study by KARABATA and HELD (1958), in connection with strontium-90 contamination of fish in the area of the Eniwetok Atoll, describes a method involving a combination of nitric acid precipitations and ion-exchange technique. In recovery experiments on fish meal ash fortified with strontium-90, yields of 100 ± four percent are reported. Four pairs of duplicate samples, at levels of activity of about 100 disintegrations per minute per gram of ash, show agreement within plus or minus five percent. Duplicate aliquots of crab skeletons, fortified with strontium-90, result in average recoveries of 99 percent with a standard deviation of plus or minus two percent. The separation of strontium from calcium is effected from 79 percent nitric acid. The reduction of the calcium-strontium ratio is sufficiently lowered so that a cation exchanger could be used effectively. The column is stripped of all remaining contaminants except strontium-89, calcium-45, and barium-140, by elution with 0.5 percent oxalic acid and five percent ammonium citrate at pH 3.5. After yttrium-90 is "grown in" completely on the column it is eluted with a small volume of ammonium citrate solution.

An evaluation by SUNDERMAN and MEINKE (1957) of the various radiochemical separation procedures applicable to the strontium-90 determination is most interesting from the standpoint of its essential agreement in the results of nitric acid separation of calcium from strontium, as reported originally by Willard and Goodspeed (1936). Using 80 percent nitric acid, Sunderman and Meinke note that 51 percent of calcium is precipitated with the strontium, and at 60 percent nitric acid, only 2.6 percent of calcium is co-precipitated with strontium. However, at 60 percent nitric acid concentrations, only 81 percent of the strontium is precipitated. Seventy percent nitric acid is considered to be a good compromise for this situation, since 98 percent of the strontium is separated under this condition. Table 3 in Sunderman and Meinke (1957) is a comprehensive summary of the extent of contamination of alkaline earth oxalates and sulfates by the activities of other fission products.

A method by Kooi (1958) for the determination of strontium-90 and strontium-89 in water is indicative of the various separation steps necessary to obtain fairly accurate results. Since recovery experiments are included in his paper, it is possible to evaluate the effects of these steps. The initial concentration step involves the addition of strontium and barium carrier to a three-liter sample of water. The carbonates of calcium, barium, and strontium are precipitated in ammoniacal solution. For the removal of calcium, a single nitric acid precipitation with 65 percent nitric acid is usually found sufficient for removing interfering elements. After this precipitation, the strontium and barium nitrates are washed with a 1:1 mixture of alcohol and ether to remove excess acid. A 70 percent recovery of strontium is attained with two precipitations. Trace experiments with calcium-45 indicate that, after two precipitations, less than 0.2 percent of calcium, originally present as one gram, was present in the nitrate precipitate. Other experiments with strontium-89 tracer indicate that, after one precipitation, the loss of strontium amounts to 10 to 12 percent and is independent of calcium concentration. Barium is removed according to the procedure of Glendenin (1951) based on the insolubility of barium chloride in concentrated hydrochloric acid-ether solution. The barium chloride precipitation is usually repeated three times. Experiments with strontium-89 indicate that its loss on the barium chloride precipitate amounts to 10 to 20 percent. A ferric hydroxide scavenge effects the removal of lanthanum-140 resulting from the decay of barium-140. The purified strontium solution is precipitated as the carbonate. The removal of lanthanum-140 from the final carbonate precipitate of strontium is evidenced by a tracer study with lanthanum-140. The overall yield of strontium in the final precipitate is about 60 percent. After counting the strontium carbonate precipitate, it is redissolved in acid, yttrium carrier is added, and yttrium hydroxide is separated. Repeated yttrium hydroxide precipitations result in a clear cut separation from traces of strontium-89 and -90. The yttrium is finally separated as the fluoride, prior to counting. Results of duplicate tests agreed within a few percent, except in one case where a deviation of ten percent was found. A maximum error of ten percent is assumed after chemical recovery factor corrections.

Methods for determining strontium-90 by the *United Kingdom Atomic Energy Authority* by Bryant *et al.* (1957 and 1959) follow the usual pattern of nitric acid separation of strontium from calcium, ferric hydroxide and barium chromate scavenge, with final precipitation of strontium as the carbonate. Yttrium is "milked out" as the hydroxide and converted to the oxalate.

A recent method which by-passes the fuming nitric acid separation is reported by OWERS and EVETT (1964). In this technique both calcium and strontium are precipitated together as the oxalate under controlled pH conditions. The reprecipitated materials contain virtually only alkaline and rare earth cations. The usual barium chromate and ferric hydroxide scavenge is resorted to. The combined oxalates are redissolved in acid and yttrium carrier is added. After the usual equilibration period of two weeks, the yttrium-90 is "milked out" as the hydroxide. After two additional reprecipitations as the hydroxide to ensure complete removal of calcium and strontium, the hydroxide is converted to the oxalate, mounted, and counted twice to check decay. After completion of counting, yttrium oxalate is ignited and the oxide residue is weighed for determination of yttrium recovery. WENDLANDT (1959) reported that from 45° to 180 °C. the hydrates of yttrium oxalate vary from $9H_2O$ to $2H_2O$. From 260° to 410 °C. the conversion from the dihydrate to the anhydrous salt takes place. The anhydrous salt is converted to the oxide from 410° to 735 °C. Essentially the same results are reported by HAHN (1963).

Strontium carrier is not added in this procedure. The natural strontium present in the sample is precipitated with calcium as the carbonate and weighed. A weighed portion is taken for counting strontium-90. Tracer strontium-85 is added at the beginning of the procedure for determination of a chemical recovery factor.

Aside from the elimination of repeated nitric acid separations, there seems to be little advantage over that of the HARLEY (1963) method. There is a distinct disadvantage due to the increased self-absorption of beta particles in the calcium carbonate precipitate. Counting efficiencies of seven percent for strontium-90 and 8.5 percent for strontium-89 indicate that very low activity samples could not be counted effectively unless large amounts of original food material is processed.

IV. Ion-exchange and other chromatographic methods

The method of PORTER and KAHN (1964) involves passing milk samples consecutively through a cation- and an anion-exchange resin column. The alkaline earth ions in milk replace the sodium ions on the cationic exchanger. The yttrium is retained as an anionic complex, probably of citrate, on the anion exchange column. The yttrium complex is broken by hydrochloric acid on the column and eluted with acid, precipitated as the oxalate, and counted. This method, modification of a previous publication by PORTER *et al.* (1961), shows improvement by solvent extraction with tributyl phosphate, after yttrium elution, to remove the lanthanum-140 daughter of barium-140. The yttrium retention by the anion exchanger is increased from 80 to 99 percent by adding sodium citrate to the milk. Elution of yttrium is increased from 81 to 90 percent by decreasing the amount of anion exchange resin and by thorough stirring of the resin during elution. The overall average yield, including other procedural manipulations, is increased to about 86 percent. The yttrium yield for combined solvent extraction (tributyl phosphate), washing, and re-extraction is 95 percent. Of the initially present lanthanum-140, 0.1 percent is found in the final yttrium precipitate. The quantitative retention of yttrium is effected on the anion-exchange column by increasing the citrate concentration of milk to 0.02 *M* from its natural previous level of 0.01 *M*.

The standard deviation of milk samples ranges from 6.0 pc/liter at levels of 13 pc/liter to 1.2 pc/liter at levels of 58 pc/liter. The 57-day yttrium-91 fission product is not separated from yttrium-90 in this procedure. Fortunately, discrimination by the cow against yttrium-91 as compared to strontium-90, is greater by a factor of 10^4.

Although not directly applied to foods, column chromatographic methods published by PIERCE *et al.* (1962 a and b), WINCHESTER (1963), and SIEKIERSKI and FIDELIS (1960) appear promising in the development of rapid column chromatographic techniques for the separation of yttrium-90 in foods. An ion-exchange method for strontium-90 in milk is decsribed by MILTON and GRUMMITT (1957), using bulk milk ash samples with the cation exchanger Dowex 50 resin and ammonium lactate as the eluant. They report yields of 85 to 95 percent for strontium-90.

V. Solvent extraction methods

The studies of PEPPARD *et al.* (1957 a and b) covering the extraction properties of tributyl phosphate and di-(2-ethylhexyl) phosphoric acid has resulted in numerous applications to the analysis of strontium-90 in foods. GOLDIN and VELTON (1961) demonstrate that yttrium is extracted by tributyl phosphate over a reasonable range of acid concentrations and that the extraction is selective with discrimination against the lighter lanthanides, specifically lanthanum. Recovery experiments with yttrium-91 indicate that triplicate determinations vary from 97 to 101 percent. The recovery of cerium is 1.5 percent and of lanthanum much less. Despite the selectivity, a number of concentration and separation steps are resorted to. Strontium is separated as the carbonate and the equivalent of ferric hydroxide scavenge is employed to remove rare earths and other polyvalent metal cations. The strontium carbonate is stored for a suitable ingrowth period of yttrium-90, dependent upon the level of activity of strontium-90 present. The carbonate precipitate is dissolved in concentrated nitric acid, and the yttrium is extracted with tributyl phosphate. After washing, stripping, and plating, the sample is counted. Since lanthanum is not extracted, the barium chromate precipitation is omitted.

The *HDEHP* method of DAVIS and ARNOLD (1964) is improved by a modification in the normality of a hydrochloric acid wash (ARNOLD and DAVIS 1965) to result in effective decontamination of the following nuclides from the final yttrium oxalate precipitate: Ru-Rh-106, Zr-Nb-95, Ce-Pr-144, Mo-99, Zn-65, Cr-51, Fe-59, Ba-La-140, Cs-137, Nd-Pm-147. The contamination of these radionuclides in the final yttrium oxalate precipitate vary from 0.01 percent to approximately 0.0001 percent. The cerium-praseodymium-144 contamination of the yttrium oxalate precipitate is reduced to approximately 0.0002 percent. All of these separations of contaminants are done by two series of extractions to result in a recovery of 100.4 ± 4.3 percent for five replicate determinations of strontium-90 and a recovery of 100.1 ± 3.5 percent for a series of six replicates. The method should be applicable to fresh, heavy-fallout food samples without resorting to long drawn out separations. Samples containing large amounts of calcium, such as bone ash, are readily adaptable to the method without resorting to repeated nitric acid precipitations.

A method used by the *Oak Ridge National Laboratory*, as described in their *Master Analytical Manual*, for the determination of strontium-89 in aqueous solutions of strontium-90, separates the yttrium-90 daughter product of strontium-90 by chelation with thenoyl trifluoracetone (TTA). The yttrium is extracted into a TTA-xylene solution. One extraction from an aqueous strontium solution at pH 4.0 to 4.5 removes 95 percent of the yttrium-90 activity. If the pH of the aqueous is 4.0 to 4.5, less than 0.5 percent of the strontium-90 is extracted into the organic phase. VOLCHOK *et al.* (1957) utilize the TTA chelation technique for bone ash. In this method yttrium carrier is added to an aliquot of dissolved bone ash solution and yttrium oxalate precipitated at pH 1.0. The precipitate is ignited, dissolved in nitric acid, and the yttrium is separated from calcium by successive ammonium hydroxide precipitations. The yttrium is extracted into a TTA-benzene solution from an aqueous solution at pH 5, back-extracted with dilute nitric acid, precipitated as oxalate, weighed to determine yield, and counted for yttrium-90 beta activity.

The chromatographic ion-exchange and solvent-extraction methods, as attempts to simplify the classical gravimetric separation steps of the earlier strontium-90 methods, are a step in the trend toward simplification. If the environmental contamination of our food supply by fallout diminishes significantly due to the cessation of atmospheric testing, it may be possible to devise more simplified techniques of analyses for the situation at hand. However, the problem of handling still larger amounts of food samples in order to attain sufficient activity for good counting statistics will result in the tendency to complicate the early stages of analysis of strontium-90 in foods.

Many other reference citations of methods have not been included in this review since they follow the same general pattern of methodology described here. In a number of instances it has not been possible to evaluate accuracy and precision due to omission of recovery experiments and standard deviation values.

VI. Conclusion

Strontium-90 determinations may be grouped into three general classifications of types of methods; liquid-liquid partition techniques appear to be most promising from the standpoint of rapidity of analysis with accuracy comparative with the classical gravimetric methods. With the cessation of atmospheric testing, it will be necessary to find simpler ways of processing still larger amounts of food material during the preliminary steps of analysis.

The determination of yttrium-90 is used in order to circumvent the difficulties involved in the presence of significant amounts of strontium-89 with strontium-90 for periods up to 300 days after atomic detonation.

Summary

Analytical methods for the determination of strontium-90 in foods have been categorized as liquid-liquid partition, ion-exchange, or gravimetric separation types of methods.

The various types of methods have been evaluated. It has been concluded that liquid-liquid partition techniques offer the most promise from the standpoint of rapidity of analysis combined with acceptable accuracy and precision.

Résumé*

Les méthodes analytiques pour le dosage du strontium 90 dans les aliments ont été classées selon qu'elles font appel à la partition liquide-liquide, l'échange d'ions ou la séparation gravimétrique.

Ces divers types de méthodes on été évaluées. La conclusion est que les techniques basées sur la partition liquide-liquide offrent le plus d'intérêt en ce qui concerne à la fois la rapidité, la fidélité et la précision.

Zusammenfassung**

Einteilung der analytischen Methoden zur Bestimmung von Strontium-90 in Lebensmitteln in die Typen der flüssig-flüssig-Verteilung, des Ionenaustausches oder der gravimetrischen Trennung.

Die verschiedenen Typen der Methoden wurden bewertet. Es wurde gefolgert, daß die flüssig-flüssig-Verteilungstechnik vom Standpunkt der Schnelligkeit der Analyse zusammen mit der annehmbaren Genauigkeit am meisten verspricht.

References

AFANASIEVA, L. I.: Use of EDTA for the separation of barium, strontium, and calcium. Zhur. Anal. Kim. **14**, 294 (1959).

ARNOLD, A., and S. DAVIS: The specificity of di(2-ethyl-hexyl) phosphoric acid (HDEHP) method for yttrium-90 in foods: Modification for fresh fallout conditions. J. Assoc. Official Agr. Chemists **48**, 5 (1965).

BRYANT, F. J., A. C. CHAMBERLAIN, A. MORGAN, and G. S. SPICER: Radiostrontium fallout in biological materials in Britain. *United Kingdom Atomic Energy Authority Research Group*, AERE HP/R2056 (1957).

—, A. MORGAN, and G. S. SPICER: The determination of radiostrontium in biological materials. AERE R-3030 (1959).

* Traduit par R. TRUHAUT.

** Übersetzt von F. BÄR.

DAVIS, S., and A. ARNOLD: The determination of strontium-90 in foods by the di-(2-ethyl-hexyl) phosphoric acid (HDEHP) method. J. Assoc. Official Agr. Chemists **47,** 580 (1964).

FARABEE, L. B.: Radiochemical analysis of strontium and barium in human urine. Amer. Med. Assoc. Arch. Ind. Health **17,** 200 (1958).

GLENDENIN, L. E.: Radiochemical studies. The fission products. Ed C. D. Coryell. Vol. 3, p. 1460. New York: McGraw Hill 1951.

GOLDIN, A. S.: *U.S. Atomic Energy Commission* document TID-7517, Part 1 b (1956).

—, and R. J. VELTEN: Application of tributyl phosphate extraction to the determination of strontium-90. Anal. Chem. **33,** 149 (1961).

—, —, and G. W. FRISHKORN: The determination of radioactive strontium. Anal. Chem. **31,** 1490 (1959).

HAHN, P.: *U.S. Public Health Service,* NE Radiological Health Laboratory, Winchester, Mass. Monthly Rep't., p. 12. April 1963.

HALLDEN, N. A., and J. H. HARLEY: Laboratory Rep't., 57-6. Anal. branch, Health and Safety Laboratory, *Atomic Energy Commission.* New York, N. Y. Nov., 20, 1957.

HARLEY, JOHN H.: Radiochemical determination of strontium-90. J. Assoc. Official Agr. Chemists **46,** 371 (1963).

Hasl Manual of Standard Procedures, NYO-4700. NY Operations Office, Health and Safety Laboratory, Anal. Div., *Atomic Energy Commission* (1959—1964).

KAWABATA, T., and E. E. HELD: A method for the determination of strontium-90 in biological materials. Contract No. AT (45-1) 540 A.E.C. Applied Fisheries Laboratory, Univ. of Washington, Seattle, Wash. Jan. 2, 1958.

KOOI, J.: Quantitative determination of strontium-89 and strontium-90 in water. Anal. Chem. **30,** 532 (1958).

MILTON, G. W., and W. E. GRUMMITT: Ion exchange methods for the quantitative separation of the alkaline earths, and their application to the determination of strontium-90 in milk ash. Can. J. Chem. **35,** 541 (1957).

MURTHY, G. K., J. E. COAKLEY, and J. E. CAMPBELL: A method for the elimination of ashing in strontium-90 determination of milk. J. Dairy Sci. **43,** 151 (1960).

Oak Ridge National Laboratory, Master Analytical Manual. *Atomic Energy Commission.* A.E.C. Method No. 9-07033802.

OWERS, M. J., and T. W. EVETT: The determination of radiostrontium in milk and herbage. Analyst **89,** 544 (1964).

PEPPARD, D. F., G. W. MASON, J. L. MAIER, and W. J. DRISCOLL: Fractional extraction of the lanthanides as their di-alkyl orthophosphates. J. Inorg. Nuclear Chem. **4,** 334 (1957 a).

—, —, and S. W. MOLINE: The use of dioctyl phosphoric acid extraction in the isolation of carrier-free yttrium-90, lanthanum-140, cerium-144, Pr-143, and Pr-144. J. Inorg. Nuclear Chem. **5,** 141 (1957 b).

PIERCE, T. B., and P. F. PECK: Use of di(2-ethyl-hexyl) orthophosphoric acid for the separation of elements, lanthanum-gadolinium, by reverse-phase partition chromatography. Nature **194,** 597 (1962).

PORTER, C., D. CAHILL, R. SCHNEIDER, P. ROBBINS, W. PERRY, and B. KAHN: Determination of strontium-90 in milk by an ion-exchange method. Anal. Chem. **33,** 1306 (1961).

—, and B. KAHN: Improved determination of strontium-90 in milk by an ion exchange method. Anal. Chem. **36,** 676 (1964).

SIEKIERSKI, S., and I. FIDELIS: Separation of some rare earths by reversed-phase partition chromatography. J. Chromatog. **4,** 60 (1960).

STEVENSON, P. C., and W. F. NERVIK: The radiochemistry of the rare earths, candium, yttrium, and actinium. NAS-NS-3020, O.T.S., *Dept. of Commerce,* Washington 25, D.C. Feb. 1961.

SUNDERMAN, D. N., and W. W. MEINKE: Evaluation of radiochemical separation procedures. Anal. Chem. **29,** 1578 (1957).

United Nations: Report of the *United Nations Scientific Committee* on the effects of atomic radiation. Gen. Assembly Official Records. 13th session. Suppl. no. 17 (A/3838) New York 1958.

U. S. Public Health Service: NERHL-64-1. Analysis of environmental samples, chemical and radiochemical procedures. NE Radiological Health Laboratory, Winchester, Mass. Apr. 1964.
— Analytical methods of the SW Radiological Health Laboratory, Las Vegas, Nev. Feb. 29, 1964.
— Procedures for the determination of stable elements and radionuclides in environmental samples. SE Radiological Health Laboratory, Montgomery, Ala. 1964.
VELTEN, R. J., and A. S. GOLDIN: Simplified determination of strontium-90. Preferential extraction of yttrium with tributyl phosphate. Anal. Chem. **33**, 128 (1961).
VOLCHOK, H. L., J. L. KULP, W. R. ECKELMANN, and J. E. GAETJEW: Determination of strontium-90 and barium-140 in bone, dairy products, vegetation, and soil. Ann. N. Y. Acad. Sci. **71**, 295 (1957).
WELFORD, G. A., and D. C. SUTTON: A rapid method for the estimation of radiostrontium in milk. 140th Nat. Meeting, Amer. Chem. Soc., New York City, Sept. 1960.
WENDLANDT, W. W.: Thermal decomposition of the rare earth metal oxalates. Anal. Chem. **31**, 408 (1959).
WILLARD, H. H., and E. W. GOODSPEED: The separation of Sr, Ba, and Pb from Ca and other metals by precipitation as nitrates. Ind. Eng. Chem., Anal. Ed. **8**, 414 (1936).
WINCHESTER, JOHN W.: Rare earth chromatography using bis-(2-ethyl-hexyl) orthophosphoric acid. J. Chromatog. **10**, 502 (1963).
World Health Organization: Tech. Rep't. Series no. 173. Methods of radiochemical analyses. Palais des Nations, Geneva 1959.

Use of ion-exchange resins in residue analysis

By

A. CALDERBANK*

Contents

I. Introduction

Ion-exchange resins are insoluble macromolecules containing fixed acidic or basic groupings. These acidic or basic groups, which are chemically bound to the insoluble matrix of the polymer, can ionize and the mobile counter ions are capable of exchanging with other ions in the surrounding solution without change in the structure of the ion-exchange material. Ion-exchange has proved to be a valuable technique for concentrating, isolating, purifying, or recovering ionic materials in solution, particularly when present in dilute quantities. The many and varied applications in organic chemistry and biochemistry have been comprehensively reviewed by CALMON and KRESSMAN (1957) and in analytical chemistry by SAMUELSON (1963). Despite the nu-

---

* Agricultural Division of Imperial Chemical Industries Ltd., Jealott's Hill, Bracknell, Berkshire, England.

merous applications in analytical chemistry, and in particular for the analysis of trace constituents, this technique has received little attention from the pesticide residue analyst.

The concentration of substances present in only minute traces often involves extremely tedious and time-consuming procedures, such as multiple extractions followed by the evaporation of large volumes of solvent. This type of concentration procedure has often to be followed by a purification, or cleanup procedure, involving the use of an inert chromatographic adsorbent. Such materials are subject to variation in properties and their reproducibility is often suspect. Ion-exchange can sometimes provide a much more convenient alternative to these methods, thus eliminating the errors and uncertainties inherent in them. This technique is ideally suited to residue analysis provided that the pesticide is ionic or can be converted into an ionic form. A large volume of the crude extract containing traces of the pesticide is passed through a column of the appropriate resin, e. g., a sulphonic acid resin if a cation is being sought, and the pesticide will be adsorbed by the resin under the right conditions, no matter how dilute is the applied extract. Elution of the pesticide with a small volume of acid or salt solution simultaneously obtains the cation in a more concentrated form for subsequent determination and effects considerable purification by eliminating naturally occurring materials not retained on (or retained on, but not eluted from) the resin. A further advantage of the use of ion-exchange resins is that their properties are reproducible, which is most important when the method must be worked in other laboratories, often in different parts of the world.

The application of cation-exchange resins for the determination of the bipyridylium herbicides, diquat [1] and paraquat, which is discussed in more detail below, is perhaps obvious. They are highly polar compounds; their salts are completely dissociated in aqueous solution and they cannot be extracted into water-immiscible organic solvents. Owing to their cationic nature, however, they are readily retained on a cation-exchange resin when dilute aqueous solutions are allowed to percolate through the resin. This is perhaps the only satisfactory technique which could be used for separating residues of these herbicides prior to analysis. Less obvious is the use of resins in the analysis of weakly basic or acidic pesticides which are soluble in organic solvents. Although ion-exchange resins are most frequently used in aqueous medium, mixed aqueous-organic or even purely organic media may be used, provided they swell the resin satisfactorily. Furthermore the solubility of the pesticide in water need not be very great for this technique to be applicable since this type of analysis is necessarily concerned with very low concentrations. A good example here is the application of ion-exchange for the determination of menazon which again is discussed in more detail below. It is a weak base (pK_a 3.8) with a solubility of about 250 p.p.m. in water at 25° C. and is retained on a cation exchange resin from an aqueous-methanol extract at pH two.

---

[1] Chemical names and structures of pesticides mentioned in text are presented in sections IV and V.

II. Ion-exchange resins

In choosing the proper resin for a particular application it is important to consider the type required (acidic or basic) and the desired ionic form, the degree of cross linkage (usually defined by the percentage of divinyl benzene [DVB] used in the copolymerisation), and the particle size (mesh size) of the resin.

a) Types of resin

The ionisable groups are the functional parts of the resin since it is these which participate in the exchange processes. They are either acidic (cation exchangers) or basic (anion exchangers). Each group may be "weak" or "strong" so that there are four main types of ion-exchange groups, viz. weak acid, strong acid, weak base, and strong base.

The first synthetic ion-exchange materials, introduced by ADAMS and HOLMES (1935), were based on the insoluble phenol-formaldehyde and aromatic amine-formaldehyde resins. These resins have been largely replaced in more recent years by materials based on polystyrene which have many advantages, including a well-defined degree of crosslinking. The degree of crosslinking, which is a most important property of the resin, is controlled by varying the proportion of divinyl benzene (DVB) present during the copolymerisation of styrene. The cross linked polystyrene matrix can then be converted into the four main types of ion-exchange resin. Thus, direct sulphonation gives a strong-acid cation exchanger containing nuclear sulphonic acid groupings. Weak cation exchangers, containing carboxyl groups, are made by first chloromethylating the polystyrene and converting this to the nitrile which is then hydrolysed. The chloromethylated product can alternatively be reacted with an amine to give an anion exchanger, either strong or weak depending on the type of amine used. The weakly basic exchangers have amino ($-NH_2$), mono-, and di-substituted amino groups, while the quaternary ammonium group, formed from tertiary amines, gives rise to strongly basic resins comparable in strength to the caustic alkalis.

The sulphonated polystyrene resins, which are the most widely used, are highly ionised in both acid or salt forms and consequently show ion-exchange behaviour over the full pH range. Adsorption of a base from solution passing through the resin results in liberation of the equivalent of acid or salt depending on whether the resin is in the hydrogen or sodium ion form. The ion-exchange reaction is similar to that occurring in ionic solution where M represents the organic matrix of the resin, e. g.,

$$(M\mathrm{SO_3})^- \left\{\begin{matrix}\mathrm{H^+}\\ \mathrm{or}\\ \mathrm{Na^+}\end{matrix}\right. + \mathrm{B^+\,Cl^-} \rightarrow (M\mathrm{SO_3})^-\,\mathrm{B^+} + \left.\begin{matrix}\mathrm{H^+}\\ \mathrm{or}\\ \mathrm{Na^+}\end{matrix}\right\}\mathrm{Cl^-}$$

The strongly basic anion-exchange resins, which all incorporate a quaternary ammonium type of structure, are also highly ionised and can be used over the entire pH range. They are usually supplied in chloride or hydroxide form and are capable of adsorbing acids from solution with liberation of the corresponding equivalent of alkali or salt, e. g.,

$$(M\mathrm{CH_2NR_3})^+ \begin{cases}\mathrm{OH^-}\\ \text{or}\\ \mathrm{Cl^-}\end{cases} + \mathrm{Na^+\,A^-} \rightarrow (M\mathrm{CH_2NR_3})^+\,\mathrm{A^-} + \mathrm{Na^+} \begin{cases}\mathrm{OH^-}\\ \text{or}\\ \mathrm{Cl^-}\end{cases}$$

The weakly acid and basic resins have more limited use. They are highly ionised only when in the salt form, the weakly acid resins being operable only in the pH range above seven and the weakly basic below pH seven. They are generally used for removing free bases or acids from solution, especially sensitive materials in buffered solution when the stronger acidic or basic resins cannot be used. The following (Table I) more commonly used resins, made by different companies, are roughly equivalent. A more comprehensive list of commercial materials is given by CALMON and KRESSMAN (1957).

Table I. *Equivalent resins*

Resin type	Dow Chemical Co.	Permutit Co. Ltd., London	Rohm & Haas Co.
Sulphonated cross-linked polystyrenes	Dowex 50 W	Zeo-Karb 225	Amberlite IR-120
Carboxylic resins	—	Zeo-Karb 226	Amberlite IRC-50
Strong base anion-exchange resins	Dowex 1	De-Acidite FF	Amberlite IRA-400

Most manufacturers further define their resins by the degree of cross-linking and mesh size. Thus Dowex 50W-X8, 100—200 mesh, or Zeo-Karb 225 (eight percent DVB), 52—100 mesh, are approximately equivalent resins containing eight percent cross-linking but of different mesh sizes.

b) Degree of cross-linking

The degree of cross-linking in a polystyrene resin refers to the percentage of DVB used in the original copolymerisation. It is of prime importance in the properties of a resin since it affects the extent of swelling, the rate of exchange, and the exchange selectivity.

The greater the degree of cross-linking the greater the number of benzene rings available for activation. Thus higher cross-linked resins have more of the active sites and consequently a greater exchange capacity. There is however, a limit to the extent of cross-linking possible or in fact desirable. Most resins are available with up to about 20 percent cross-linkage and it is difficult to prepare them with a higher proportion than this. The very highly cross-linked resins have also limited use because of their much reduced swelling characteristics.

Because ion-exchangers contain polar groupings they swell in water or other polar liquids and the higher the degree of cross-linking the lower the degree of swelling. The extent of swelling affects the rate of exchange since the diffusion of ions within the resin is faster when the resin is highly swollen. Thus higher cross-linked resins give slower equilibration rates especially where large ions are concerned and consequently a lower degree of cross-linking is desirable for the exchange of organic ions. The slower equilibration rate with the higher cross-linked resins is often shown by "tailing" during the elution stage. This is illustrated in the method developed by CALDERBANK and TURNER (1962) for menazon where a one percent cross-linked resin was used in preference to eight percent because of the much sharper elution peak obtained with the former.

The lower cross-linked ion-exchangers have some minor disadvantages in practice. Because of their greater swelling characteristics they occupy a greater bed-volume for a given exchange capacity and are thus not so economical in space. The swollen resin also shrinks to a marked extent in the presence of more concentrated solutions of ions, which occurs at the adsorption and particularly at the elution stage. This can be inconvenient as it may disturb the packing of the resin column. A more serious disadvantage of the lower cross-linked resins is that they have very little selectivity. This is particularly important in the case of chromatographic separations which may depend on small differences in affinity. In residue analysis one is concerned with collecting or retaining a particular organic ion as selectively as possible. The procedure usually involves passing a crude extract of plant or animal origin through the ion-exchange column. The lower cross-linked resins are much more indiscriminate in adsorbing considerably more of the naturally occurring ions present, often at the expense of the pesticide. The result is that adsorption of the pesticide may not be complete and furthermore the final ion-exchange effluent may contain a considerable amount of naturally occurring material which may interfere with the final determination of the compound. Thus in selecting a resin with the appropriate degree of cross-linking one has often to reach a compromise between a high exchange rate with poor selectivity which requires low cross-linking and the lower exchange rate but greater selectivity of the higher cross-linked resins. The right selection can often only be made on the results of experiments.

c) Mesh size

Ion-exchange resins are manufactured in the form of spherical beads ranging from about 0.05 to 1.5 mm. in diameter, the particle size being established on standard sieves and specified according to the mesh size.

The exchange rate is obviously faster with the finer resins, down to 200 or 400 mesh, but a limiting factor is the restricted flow rate through these columns. In general, the finer grades of resin are used for chromatographic separations whilst the larger diameter beads are suitable for simple concentration or replacement techniques. Since the rates of exchange with large organic ions are low, an exchanger with fine particle size and a low degree of cross-linking may be necessary. The 50—100 mesh size resin is a most useful, general purpose grade allowing reasonably fast flow rates with quite good exchange efficiency.

III. Experimental procedures

Since ion-exchange resins are supplied in the form of spherical beads they are well-suited to column technique which is standard practice in nearly all ion-exchange work (SALMON and HALE 1959). The column of resin can be conveniently supported on a half-inch plug of glass wool at the bottom of a burette or glass tube fitted with a stopcock. There is no difficulty in packing the column. A slurry of the resin is simply poured into the tube, allowed to settle, and then either used directly or, if necessary, converted into the desired ionic form by allowing a solution of acid or salt to flow through it. The only precaution which need be taken is to ensure that the column does not dry out but is kept covered with water or solution at all times. Up to a dozen columns can be handled conveniently at any one time so that once the technique has been established it is possible to streamline it to carry out a series of determinations simultaneously. By the use of constant flow devices, in which the solution (or effluent) emerges at a point higher than the top of the bed of resin, it is possible to leave a series of columns, with extracts passing through them, without any attention whatsoever.

After the most suitable resin has been selected for a particular application experiments must be carried out with the pesticide in pure solution to determine the most convenient mesh size and cross-linkage and also the best elutant. An elutant giving a good recovery of the pesticide in the smallest volume is to be preferred, although a further factor to be considered is that the elutant should not interfere in the method used for final determination of the pesticide.

The amount of resin, or height of the resin bed required, cannot be simply arrived at from a knowledge of the exchange capacity of the resin. The amount of naturally occurring substances present in a plant extract is likely to be an unknown and variable factor and consequently it is necessary to carry out recovery tests with the pesticide in the presence of plant extract using varying amounts of resin.

In most of the applications of ion-exchange, the columns of resin can be used over and over again by regeneration after use with the appropriate acid or salt solution. This is not true of residue analysis, where the extracts passing through the columns are likely to contain high concentrations of plant pigments and organic ions which may be difficult to remove from the resin. It is common practice to reject physical adsorbents after they have been used for cleanup operations in pesticide residue analysis and the same is true of ion-exchange resins. The amount of resin involved in a single determination is small and consequently it is much more convenient to start with a fresh column of new resin for each determination.

The rate of passage through the column is controlled by the flow resistance of the resin bed and may be further regulated by partially closing the outlet tap or stopcock. The efficiency of adsorption is less at fast flow rates but in residue analysis faster flow rates can often be used with advantage. In this type of analysis one is concerned with the adsorption of traces of an organic compound in the presence of a considerably larger quantity of background or naturally occurring substances which may be retained to a greater or lesser extent by the ion-exchange resin. In practice it is more convenient to use the fastest flow rate at which the pesticide is retained with a minimum of interfering substances. The same principle applies to the elution step, although an increase in the elution rate over a certain value is not efficient and only increases the total volume of effluent.

After the adsorption stage, the resin column can often be conveniently washed with a solution of acid or salt which selectively removes some of the more easily eluted naturally occurring substances which might otherwise interfere with the final determination of the pesticide.

Various factors must be taken into account in selecting the best eluting agent for recovering the pesticide from the resin. Thus one might expect to elute more efficiently, or collect the pesticide in a smaller volume of effluent, using a solution containing ions which have a higher affinity for the exchanger. The multivalent inorganic ions have generally higher affinity for a sulphonic exchanger, e. g., $Al^{+++} > Ca^{++} > K^{+} = NH_4^{+} > Na^{+} > H^{+}$. There may, however, be reasons why multivalent ions cannot be used and it is often more convenient to obtain sharper elution peaks simply by increasing the concentration of the salt solution or acid used. In the recovery of weak bases from a cation-exchange resin the use of an alkaline eluting agent, e.g., ammonium hydroxide, is an advantage, since the dissociation of the organic base is then suppressed. There is a further advantage to using ammonium hydroxide as an elutant. The solution of pesticide recovered from the column in ammonium hydroxide can be readily concentrated by evaporation avoiding the presence of high concentrations of unwanted electrolytes. In many cases the recovery of a basic compound from a cation-exchange resin is facilitated if a water-miscible organic solvent, such as methyl alcohol is added to the eluting agent.

Similar procedures may be used for the recovery of organic acids from anion-exchange resins. Mineral acid is most commonly used with the strongly basic resins and here, of course, the anion is the effective eluting agent.

The temperature of the column during elution also has a profound effect on elution rates and in many instances the volume of elutant required can be reduced if a warm solution is used. Elution is a relatively slow process and the solution will soon cool unless the column is provided with some form of external heating. The mechanical difficulty in setting up and using heated columns on a routine basis usually precludes this technique in practice.

The method used finally to determine the pesticide depends upon the compound to be analysed. The actual determination can be carried out on an aliquot of the ion-exchange effluent directly (e.g., bipyridylium herbicides), after concentration by evaporation (e.g., amitrole), or after extraction with an organic solvent (e.g., menazon). It is common practice to wash the ion-exchange resins before use with a solution of the salt used as elutant. This has the advantage of removing traces of impurities from the resin which might otherwise have been eluted with the pesticide and interfered with its determination.

IV. Applications in residue analysis

a) Amitrole

RACUSEN (1958) first used an ion-exchange resin to cleanup plant extracts containing the herbicide amitrole (3-amino-1,2,4-triazole) and its metabolites prior to paper chromatography.

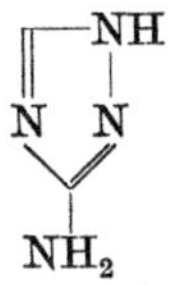

Amitrole

The plant material was extracted with water and the homogenate heated for a few minutes to precipitate protein. The clear filtered extract was then passed through a column of the cation exchanger, Amberlite IR 120 H, contained in a 50-ml. burette. The basic herbicide was retained by the resin, which was washed with water before elution using 3*N* ammonium hydroxide solution.

This technique was developed into a method for determining residues of amitrole in plant material by HERRETT and LINCK (1961 a and b). The ion-exchange effluent containing the purified amitrole was concentrated to near dryness, made up to standard volume, and the amitrole determined colorimetrically. Interference from background absorption of control samples was negligible at the wavelength (538 mμ) at which measurements were made.

In a method described by STORHERR and BURKE (1960) the plant material was extracted with aqueous methanol and the amitrole adsorbed on Dowex 50 resin.

b) Bipyridylium herbicides

Diquat (1,1′-ethylene-2,2′-bipyridylium) dibromide and paraquat (1,1′-dimethyl-4,4′-bipyridylium) dichloride or dimethylsulphate are the most active members of the important new group of bipyridylium herbicides whose properties have been reviewed by CALDERBANK and CROWDY (1962 b) and more recently by SPRINGETT (1965). They are quick acting, non residual, and generally nonselective in action. A new member of this group, morfamquat [1,1′-bis(3,5-dimethylmorpholinocarbonylmethyl)-4,4′-bipyridylium] dichloride and originally coded PP 745 and described by FOX (1964 a), is selective in action and has been introduced for the post-emergence control of broad-leaved weeds in cereals. The results of trials carried out with morfamquat in 1963 and 1964 are described by FOX and BEECH (1964 b).

Although these substances show interesting differences in biological effect, their chemical and physical properties are very similar. They are highly polar compounds, are very soluble in water, and are insoluble in the water-immiscible organic solvents. Because of their cationic nature they are readily retained from dilute aqueous solutions by cation-exchange resins and this is the most convenient method of concentrating and purifying residues of these herbicides prior to determination. Although these compounds are so closely related chemically the best conditions required for adsorption to, and elution from, the resin column differ with each herbicide and consequently it is more convenient to consider the individual problems associated with the analysis of each separately.

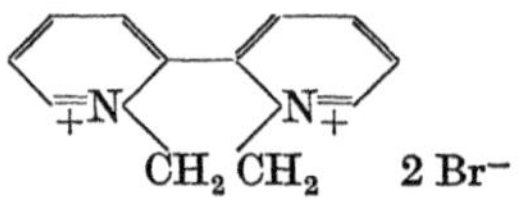

Diquat dibromide

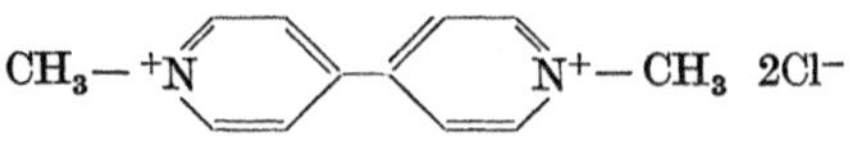

Paraquat dichloride

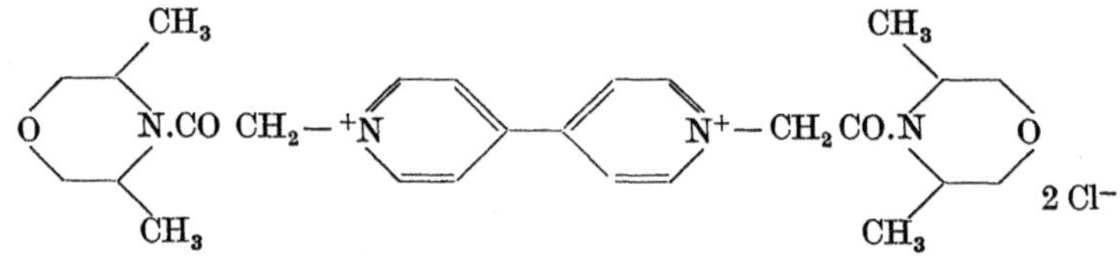

Morfamquat dichloride

1. Diquat. — Diquat is the active ingredient of "Reglone", a quick-acting desiccant, first introduced into the United Kingdom for destroying potato haulm. A specific method for determining diquat residues in potato tubers, which depends on measurement of the light absorption of reduced solutions of diquat after concentration and purification by cation-exchange chromatography, is described by CALDERBANK *et al.* (1961 a). According to this method the starch in the tubers is first hydrolysed with acid in order to release adsorbed diquat and the hydrolysate is filtered to give a clear extract. The extract is then passed through a column of the cation-exchange resin Zeo-Karb 225 (eight percent DVB, 52—100 mesh) which retains the diquat but passes most of the plant constituents. The resin is washed with 2*N* hydrochloric acid solution which removes some of the remaining plant constituents without displacing diquat. The diquat is finally eluted with a small volume of 6*M* sodium chloride solution and determined by measuring the light absorption of an aliquot of the effluent in the region of 379 mμ after reduction with sodium dithionite. Since this method was developed diquat has found increasing usage on a variety of other crops and the operating variables in the method have been subject to more critical examination in these laboratories. The method is now applicable with minor variations to all crops so far examined and also to animal tissue and excreta, milk, and water.

Zeo-Karb 225 (eight percent DVB, 52—100 mesh) resin was found to be the best of the complete range of Permutit cation exchange resins examined. With a lower cross-linked resin, e. g., Zeo-Karb 225 (one percent DVB), adsorption of diquat from pure solution is quantitative and sharper elution peaks are obtained. Difficulties, however, arise in the presence of plant extract. The latter resin gives considerable background absorption in the final effluent so that it is impossible to determine the amount of diquat recovered. Furthermore diquat is more easily removed from this resin and consequently it is not possible to carry out a selective washing procedure to remove plant constituents prior to elution of diquat.

Traces of diquat are held quite strongly to plant material and it has been found necessary in all cases to boil the macerated plant material with dilute sulphuric acid for several hours in order to release the diquat into solution. If the filtered acid hydrolysate is passed directly, without further treatment, through the Zeo-Karb 225 resin recovery of diquat in the final effluent is low and the background absorption is high. This is because the resin adsorbs large quantities of coloured plant materials at low pH and these are then eluted along with the diquat and increase the general background absorption. A better recovery of diquat is obtained if the extract is first made alkaline before passing through the resin and the background absorption in the final effluent is considerably lower. Presumably this is because ionisation of the weakly basic materials present in the hydrolysate are suppressed under alkaline conditions and consequently they are not adsorbed to the same extent on the resin column leaving more sites available for adsorption of diquat, the ionisation of which is not appreciably affected by change of pH.

In the method described by CALDERBANK *et al.* (1961) 6*M* sodium chloride solution was chosen as the most suitable elutant for displacing diquat from the selected resin. Reduction of diquat could be carried out in 6*M* sodium chloride solution as effectively as in pure solution and this was one of the main reasons for its selection. It suffers, however, from the disadvantage of producing tailing so that recovery of diquat from the resin is incomplete. Thus in order to retain sensitivity the volume of effluent collected is restricted to 25 ml. so that recovery of diquat in this volume is only 80 percent. Hydrochloric acid (5.5*N*) and ammonium chloride solution (5*M*) produce sharper elution peaks, the latter giving 95 percent recovery in 25 ml. and complete recovery in 50 ml. of effluent, but both introduce complications at the reduction stage. By modifying conditions of the reduction it was found possible to obtain reproducible results in determining diquat in 5*M* ammonium chloride solution and consequently this has been adopted as the preferred elutant.

The selection of the particular mesh size of resin and amount used in the column calls for further comment. Resin of size 52—100 mesh is recommended for general analytical use and was selected as a convenient compromise between the lower capacity and fast running of the coarser grades and the higher capacity but slower running of the resins graded with mesh sizes above 100. Resins with mesh sizes above 200 give very slow flow rates and are unsuitable for routine use. The 100—200 mesh size of Zeo-Karb 225 (eight percent DVB) gave elution peaks similar to the equivalent 52—100 mesh resin but suffered from two slight disadvantages. Passage of solution through the columns of resin was slower in the 25-ml. burettes used, necessitating the use of air pressure towards the end of the flow and rather higher background absorption was obtained in the final effluent containing diquat.

When diquat is adsorbed on five ml. of settled Zeo-Karb 225 resin (eight percent DVB, 52—100 mesh) from pure solution, recovery in 50 ml. of 5*M* ammonium chloride solution is quantitative. When diquat is added to a clear, filtered plant hydrolysate, which is then allowed to pass through the resin, the recovery in the ammonium chloride effluent is often not complete. This suggests that further loss of diquat occurs owing to non-quantitative retention by the resin in the presence of a relatively large volume of plant material hydrolysate. The cationic naturally occurring substances possibly compete with diquat for negative sites on the resin. This is not a serious difficulty as losses are usually in the order of only ten percent or so. Increasing the amount of resin would probably help in giving higher recoveries but a greater quantity of naturally occurring plant substances would also be retained and eluted along with the diquat, giving higher background absorption in the effluent. The sensitivity may also be reduced because a rather larger volume of elutant may be required to displace the diquat completely.

2. Paraquat. — A specific method for determining residues of paraquat, the active ingredient of the commercial formulation "Gramoxone", in food crops has recently been described by Calderbank and Yuen (1965). Paraquat salts, like those of diquat, are extremely soluble in water and the active cation is readily retained when dilute solutions of the salt are allowed to percolate through a cation-exchange column, thus affording a convenient means of separating paraquat from plant constituents which are largely not retained and which would otherwise interfere with the subsequent determination. The method is thus similar to that described for diquat and depends on extraction of residues by boiling the food crop with dilute sulphuric acid. The filtered extract is diluted further and passed through a column of Zeo-Karb 225 (eight percent DVB, 52—100 mesh) cation-exchange resin which retains the paraquat but allows most of the coloured plant constituents to pass through. The resin column is washed with dilute hydrochloric acid and dilute ammonium chloride solution which removes further plant constituents without loss of paraquat. The latter is finally eluted with saturated ammonium chloride solution and determined absorptiometrically after reduction with alkaline sodium dithionite solution.

As with diquat, the best of the Permutit resins for this determination was found to be Zeo-Karb 225 containing eight percent divinylbenzene. Although a good recovery of paraquat from pure solution was obtained using the equivalent one percent cross-linked resin, the recovery in the presence of plant extract was poor. The background absorption due to naturally occurring plant constituents was also higher with this resin. With a Zeo-Karb 225 resin containing a higher proportion (12 or 20 percent) of DVB, recovery of paraquat was very poor even from pure solution because of excessive tailing during elution. By using the carboxylic acid resins based on polymethacrylic acid, Zeo-Karb 226 (containing 2.5 or 4.5 percent DVB) recovery of paraquat from pure solution was good. However, the binding of paraquat to the carboxylic acid sites is much weaker and the herbicide is too easily removed from this resin, e.g., by washing the resin with 2*N* hydrochloric acid solution, a procedure necessary to reduce the background absorption due to naturally occurring plant substances in the final effluent containing paraquat. Five-ml. settled columns of Zeo-Karb 225 (eight percent DVB) resin can be washed with up to 120 ml. of 2*N* hydrochloric acid solution without appreciable loss of adsorbed paraquat.

In order to obtain acceptable recoveries of diquat from sulphuric acid hydrolysates of food crops it is necessary to neutralize the extract before passage through the resin column. Paraquat is more firmly bound to cation-exchange resins than is diquat and is retained equally well from acid extracts, so that neutralization prior to the ion-exchange stage is unnecessary. Although neutralization of the extract has the advantage of giving final ammonium chloride effluents having lower background absorption, it is not so important with paraquat. This is because reduced paraquat has its maximum absorption at a longer wavelength (396 mμ) than reduced diquat (379 mμ) and interference from the absorption of natural plant constituents is less at the longer wavelength. To minimise the high background absorption in ammonium chloride effluents, resulting from the use of acid extracts, the resin column after percolation is washed successively with 2*N* hydrochloric acid solution and then with dilute ammonium chloride solution before elution of the paraquat with saturated ammonium chloride. The final effluent thus obtained is colourless or faintly yellowish and the reduced solution does not absorb appreciably in the region of 396 mμ.

Saturated ammonium chloride solution was found to be the best elutant for displacing paraquat from Zeo-Karb 225 (eight percent DVB) and quantitative recovery could be obtained from five ml. of settled resin in a 25-ml. burette using 50 ml. of elutant. The 6*M* sodium chloride solution produced considerable tailing, giving only 50 to 60 percent recovery in the same volume of effluent.

The above method has been found to be applicable with minor modifications to all food crops so far examined and also to animal tissue and excreta, to milk, and to water. The method has been modified by Pack (1964) to give, in some cases, improved recoveries. Dowex A.G. 50W X-8 (100—200 mesh) cation-exchange resin replaces Zeo-Karb 225 (eight percent DVB, 52—100 mesh) and double the volume (i.e., ten ml.) of settled resin in a 50-ml. burette is used. Over 80 percent recoveries are obtained by using a rather slower elution rate and reducing the background absorption by first making the plant hydrolysate alkaline before percolating it through the resin column.

An interesting modification, due to Ospenson and Pack (1964), is for the determination of paraquat residues in soil. Residues of paraquat are adsorbed irreversibly to the clay mineral fraction of the soil and cannot be displaced by the simple elution techniques as used with cation-exchange resins. Even saturated solutions of ammonium, calcium, and aluminium salts cannot displace paraquat from these firm adsorption sites. Paraquat can, however, be released from the soil by refluxing it with fairly strong sulphuric acid, and in some cases five hours' refluxing with 18*N* sulphuric acid is required. This drastic treatment, which completely breaks down the soil structure, does not affect paraquat which is recovered from the extract by filtering, diluting to approximately 1*N*, and passing it through the cation-exchange resin in the usual way.

Although diquat and paraquat are most conveniently determined absorptiometrically by reduction to their respective coloured free radical, an alternative method which can also be applied directly to the ion-exchange effluent is by oscillographic polarography. Furthermore, diquat and paraquat give reduction waves at potentials sufficiently separated to suggest they could each be determined accurately in mixtures of both. DAVIES (1964), in a recent review of advances in polarography, has briefly referred to the determination of diquat and paraquat in mixtures.

3. Morfamquat. — This new bipyridyl herbicide is selective in action and has been introduced for the control of certain broad-leaved weeds in cereals, particularly corn marigold and mayweed, which are resistant to the hormone weedkillers. The development of analytical methods for determining residues of morfamquat in cereals and in soil posed a number of interesting problems which have been largely overcome by work carried out in these laboratories.

Solutions of morfamquat, like those of paraquat, form a deep blue colour on reduction due to free radical formation, and the wavelength of maximum absorption is at 398 mμ. The main problems in developing a residue method lay in extracting and concentrating the herbicide by ion-exchange prior to its determination. Like the other bipyridyls it is readily adsorbed and is difficult to remove from plant constituents, such as polysaccharides, protein, and even leaf surfaces. Prolonged boiling with dilute sulphuric acid releases the herbicide into solution but also causes some hydrolysis to the diacid (I). By using more dilute acid (0.5 *N*), and a shorter (three hour) reflux period, hydrolysis of morfamquat is not appreciable.

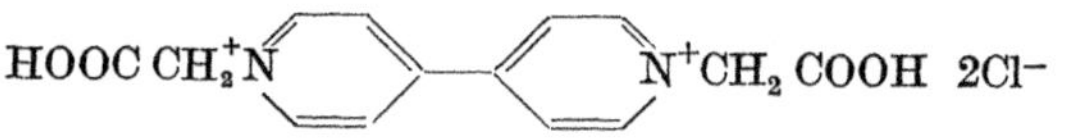

(I)

Morfamquat is readily retained from dilute solutions by most of the cation-exchange resins but considerable tailing occurs on elution from those containing a high degree of cross-linking. This is probably due to the larger size of the morfamquat molecule. Thus only 20 percent recovery of morfamquat is obtained from eight percent cross-linked Zeo-Karb 225 resin using 50 ml. of saturated ammonium chloride elutant. Progressively higher recoveries are obtained with the lower cross-linked Zeo-Karb 225 resins, the best results being obtained with a resin of two percent cross-linking, which gives 92 percent recovery in the same volume of effluent. Difficulties arise with the one percent cross-linked resin because of the higher irrelevant background absorption in the final effluent and also because morfamquat is more easily removed from this resin with the dilute ammonium chloride solution which is needed to wash the resin prior to elution of the herbicide.

We have been recently concerned with developing a method for determining residues of morfamquat in soil. Like the other bipyridyls it is extremely strongly adsorbed and cannot be removed by simple elution techniques. Prolonged boiling in strong sulphuric acid, which is necessary to release the herbicide into solution, also causes hydrolysis to the diacid (I). The problem then resolves itself into determining this compound in the presence of the soil extract.

The diacid (I) is less firmly adsorbed to cation-exchange resins than is morfamquat or paraquat and larger amounts of resin are needed to obtain acceptable recoveries. Tailing also occurs during elution of the compound with saturated ammonium chloride solution so that larger volumes of elutant are required and the sensitivity is correspondingly reduced. The higher cross-linked Zeo-Karb 225 resins retain the compound more efficiently during the adsorption stage but require a greater volume of ammonium chloride for satisfactory elution. A reasonable compromise between these two competing factors of high retentive power and ease of release is provided by the 4.5 percent cross-linked Zeo-Karb 225 cation-exchange resin. The diacid (I) forms a free radical on reduction and can be determined in ion-exchange effluents by reduction with alkaline sodium dithionite in the same manner as the other bipyridyl herbicides. Solutions of the free radical exhibit a sharp absorption maximum at 400 mμ and measurements are made at this wavelength.

c) *N-Methyl-isonicotinic acid*

SLADE (1965 a) has shown that solutions of paraquat salts decompose on irradiation with ultraviolet light in the presence of oxygen and that the quaternary salt (II) of isonicotinic acid and methylamine (III) are the principal products, thus —

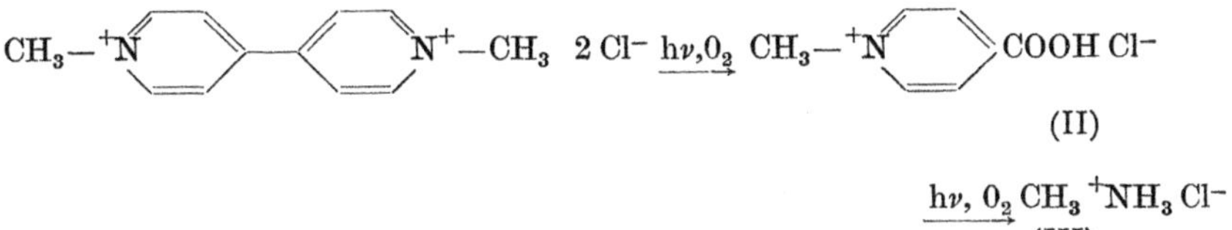

(II)

(III)

It has further been demonstrated by SLADE (1965 b) that the same products are formed by photochemical decomposition of paraquat on the surfaces of plants. This decomposition occurs slowly in strong sunlight, largely after the plants are dead, and there is no possibility of the products being translocated to seed or edible parts of the crop. However, succeeding crops grown in the same soil may pick up residues of II, and although this compound has a very low order of toxicity to mammals (oral LD_{50} to rats $>$ 5 g./kg.), a method of analysing for residues in food crops was sought.

N-Methyl isonicotinic acid can be extracted from plant material by boiling it with dilute sulphuric acid. Residues of this chemical are then recovered by allowing the extract to pass through a column of the cation-exchange resin Zeo-Karb 225 (eight percent DVB, 52—100 mesh) under the same conditions used to retain residues of paraquat. *N*-Methyl isonicotinic acid is more easily removed from the resin than paraquat and can be eluted with 1*M* ammonium chloride solution. The concentration of II in the ion-exchange effluent is then determined by oscillographic polarography at a peak potential of −0.85 to −0.90 volt relative to an internal mercury pool.

d) *Beta-Hydroxyethylhydrazine*

This is a growth regulator which has been used to induce flowering of pineapple plants and thus to control the date of harvest of the fruit. A sensitive method of analysis for determining residues of *beta*-hydroxyethylhydrazine (IV) in pineapples, relying on concentration and purification with ion-exchange resins, has been described by THOMAS and ACKERMANN (1964).

$$HOCH_2CH_2NH\,NH_2$$
(IV)

An aqueous extract of the fruit is passed successively through an anion- and a cation-exchange resin. The chemical is held by the cation-exchange resin and recovered by eluting it with dilute hydrochloric acid. An aliquot of the acid effluent containing IV is then reacted with cinnamaldehyde to form a yellow-coloured hydrazone which is measured at 420 mμ.

Dowex 1-X8 and Dowex 50W-X8 (50—100 mesh) were the resins employed in this method. The purpose of the anion-exchange resin Dowex 1, used in the OH^- form, was to remove some of the interfering pigments in the extract and 80 g. of this resin was used in each determination. The growth regulator (IV) is strongly basic and thus passes through this resin. Twenty g. of the cation-exchange resin Dowex 50 is used to retain (IV) which is eluted with 100 ml. of hydrochloric acid, leaving behind more of the interfering pigments. The method can detect down to 0.015 p.p.m. of the chemical (IV) in pineapples with an average recovery of 79 percent. The authors recommend removing the fines from the resins and also washing them immediately before use because they give a yellow effluent after they have been left standing for some time.

e) *Menazon*

Menazon [O,O-dimethyl *S*-(4,6-diamino-1,3,5-triazin-2-yl)methyl phosphorodithioate] is a selective aphicide with low mammalian toxicity and has been introduced for the control of aphids on a wide variety of fruit and vegetable crops (CALDERBANK *et al.* 1961 b).

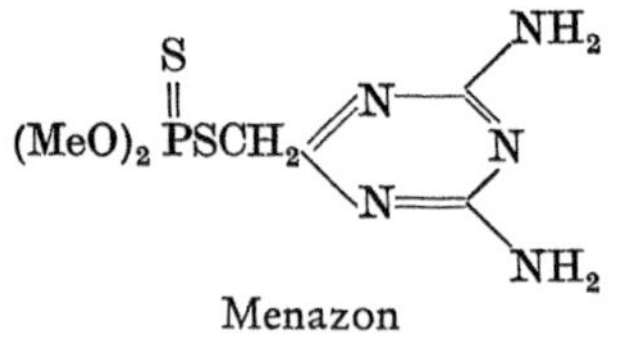

Menazon

An ion-exchange method for determining residues of menazon in food crops is described by CALDERBANK and TURNER (1962 a). The method depends on separating the insecticide from all naturally occurring phosphorus compounds, using a combination of ion-exchange and solvent extraction, before determining it by phosphate analysis. The limit of detection is of the order of 0.05 p.p.m. A modification to the final phosphate determination, which renders the method highly specific, utilizes a paper chromatographic technique.

Menazon possesses the diamino-triazine grouping and is consequently weakly basic (pK_a 3.8). It is 97 percent in the cationic form at pH two and would be expected to adhere more strongly to cation-exchange resins at low pH. Thus, although menazon is retained by the cation exchange resin Zeo-Karb 225 (one or eight percent cross-linked) equally well from neutral or acid pure solution, good recoveries are only obtained in the presence of plant extract if the extract is first acidified before passing it through the resin. The one percent cross-linked resin is used in preference to eight percent because of the much sharper elution peaks obtained with the former. Elution from eight percent cross-linked resin is accompanied by considerable tailing even when high concentrations of eluting agents are used. Ammonium acetate buffered at about pH five, is the best eluting agent. At this pH menazon is largely unionised and more readily removed from the resin.

Residues of menazon are extracted from the crop with aqueous methanol and the clear filtered extract is acidified and passed through a column of Zeo-Karb 225 (one percent DVB, 52—100 mesh) which retains the menazon but passes most of the naturally occurring phosphorus-containing compounds. The resin column is then washed with very dilute hydrochloric acid before eluting the menazon with 1*M* ammonium acetate solution at pH five in a volume of 75 ml. For many plant materials no extraneous phosphorus-containing material appears in the ammonium acetate effluent whilst with others, notably brussels sprouts and strawberries, traces of phosphorus can often be detected in the effluent. Menazon, however, has a 4.5 : 1 partition ratio in favour of chloroform and can be readily extracted from the aqueous effluent. Extraction with chloroform thus provides a further purification step and confers the added advantage that the aphicide is dissolved in a low-boiling organic solvent which can be evaporated easily. Menazon is then determined by degradation to inorganic phosphate and developing the polyphosphomolybdate blue colour, either directly or more specifically after paper chromatography of the concentrated chloroform extract.

An alternative method of determining menazon is by oscillographic polarography and this technique can sometimes be applied to the effluent from the ion-exchange column, e. g., in the case of tobacco. In this procedure residues of menazon are eluted from the resin with 95 ml. of 0.1*M* ammonium acetate solution and 5.0 ml. of concentrated hydrochloric acid is added to the effluent. The menazon is determined differentially using a blank ammonium acetate-hydrochloric acid solution in the second cell. The peak for menazon occurs in this solution at -0.53 to -0.58 volt relative to an internal mercury pool.

V. Applications in metabolism studies

Although the main emphasis of the present article has been on the use of ion-exchange resins in routine methods of residue analysis it is perhaps fitting to include here a brief mention of the uses to which this technique has been put in separating and identifiying metabolites of pesticides. Most of this work has been concerned with the organophosphorus insecticides.

PLAPP and CASIDA (1958 a) described a general technique for separating and identifying all the various phosphoric, phosphorothioic, and their mono- and dialkyl esters which are produced by hydrolytic cleavage of the well-known organophosphorus insecticides. The separations were achieved by anion-exchange chromatography on Dowex 1-X8 (100—200 mesh). Phosphoric, phosphorothioic, and mono- and dialkyl phosphoric acids were progressively eluted with higher concentrations of hydrochloric acid. Methanol and acetone were used as co-solvents to elute dialkyl phosphorothioic and dithioic acids. The order of elution of the phosphoric acid derivatives from the anion-exchange column depended on both the primary pK value and the effect of solvent action. Elution of phosphoric and monoalkyl and dialkyl phosphoric acids occurred in the order of their increasing acid strengths.

This technique has been applied to studies of the hydrolysis of P-32 labelled insecticides administered to plants, animals, and insects. The eluted hydrolysis products, located in the various fractions by radioactivity measurements, were then identified by co-chromatography with known derivatives or by infrared spectra.

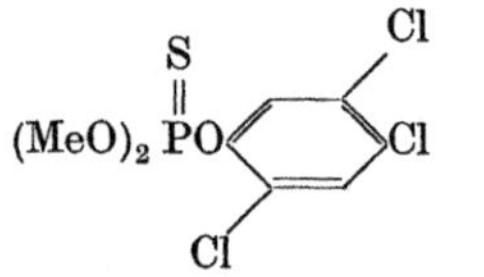

Trolene

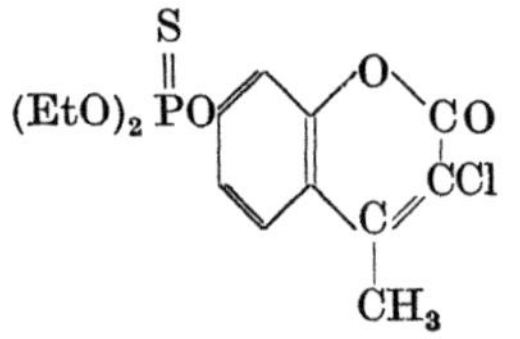

Co-Ral

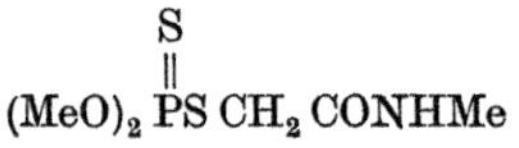

Dimethoate (Rogor)

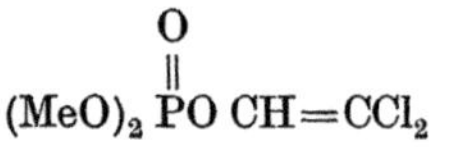

DDVP

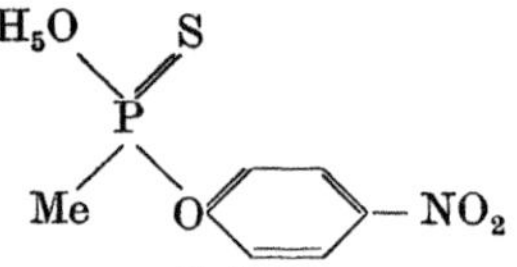

Colep

Details of the separation and identification of the mammalian metabolites of trolene [O,O-dimethyl O-(2,4,5-trichlorophenyl) phosphorothioate] are given by PLAPP and CASIDA (1958 b). Ion-exchange chromatography allowed separation of phosphoric, phosphorothioic, methylphosphoric, dimethyl phosphoric, dimethyl phosphorothioic, O-methyl O-(2,4,5-trichlorophenyl) phosphoric, and the corresponding phenyl phosphorothioic acids. Similar studies are described by KRUEGER *et al.* (1959) for Co-Ral [O-(3-chloro-4-methylumbelliferone) O,O-diethyl phosphorothioate] following dermal application to rats, a goat, and a cow, and by DAUTERMAN *et al.* (1958) for dimethoate [O,O-dimethyl *S*-(*N*-methylcarbamoylmethyl) phosphorodithioate] after oral administration to rats and cows. Ion-exchange chromatography was also applied to the separation of plant metabolites of dimethoate (DAUTERMAN *et al.* 1960) and to separate the hydrolysis products of DDVP (2,2-dichlorovinyl dimethyl phosphate) formed by mammalian tissue homogenates and plasma (HODGSON and CASIDA 1962). Both types of resin were used in a study of the metabolism of Colep [O-phenyl-

O'-(4-nitrophenyl) methylphosphonothionate] in plants and animals, described by MARCO and JAWORSKI (1964). Separation was facilitated by fractionation of the extracts on an anion-exchange (Rexyn AG 1, 40—100 mesh) and a cation-exchange (Rexyn AG 50, 40—100 mesh) resin. Phenolic metabolites present in urine were fractionated on a weak cation-exchange resin (Rexyn CG 51, 100—200 mesh) and a weak anion exchanger (Rexyn CG 8, 100—200 mesh).

In our own laboratory the des-methyl derivative (V, $X=S$) of menazon (CALDERBANK 1966) has been separated from its corresponding oxygen analogue (V, $X=O$) by linear gradient elution from the anion-exchange resin De-acidite FF (100—200 mesh, OH^- form). A mixture (160 mg.) of the two compounds was dissolved in dilute ammonium hydroxide (200 ml., pH ten) solution and the solution was passed trough a 1.5×3.5 cm. column of the resin. Both compounds were retained and separated by gradient elution with hydrochloric acid, the strength of the acid increasing from $0.001N$ to $0.136N$. The oxygen analogue (V, $X=O$) was eluted first.

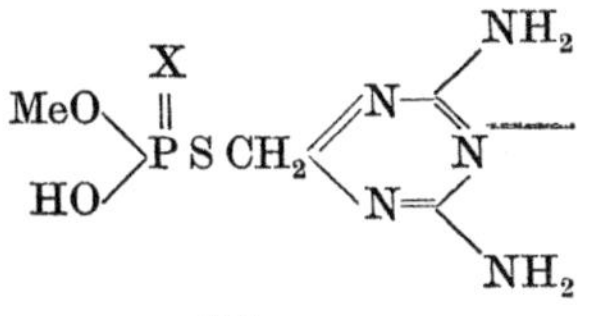

(V)

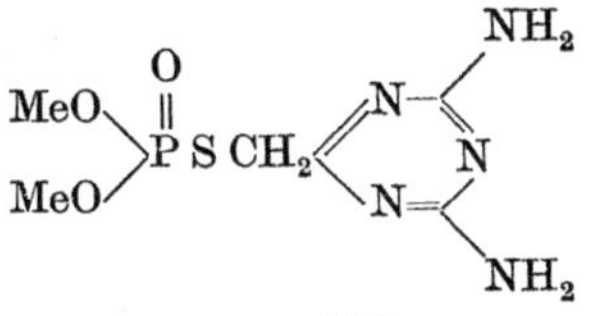

(VI)

A further interesting separation, carried out in these laboratories, was that of menazon from its corresponding thiolate VI. Both compounds adhere to the cation-exchange resin Zeo-Karb 225 (one percent DVB, 52—100 mesh) but the thiolate VI is more readily eluted and can be preferentially displaced from the resin using 0.01 M ammonium formate solution.

The use of the cation exchanger Amberlite IR-120 to clean up plant extracts containing the herbicide amitrole and its metabolites (RACUSEN 1958) has been further elaborated by HERRETT and BAGLEY (1964) and a third, herbicidally active metabolite has been detected.

VI. Conclusions

Provided the pesticide is ionic or can be converted to an ionic form, ion-exchange resins offer a useful method of concentrating and purifying traces of pest-control chemicals prior to analysis. The columns need little attention in operation and by using a series of columns it is possible to carry out up to a dozen determinations simultaneously. Analysis can be carried out spectrophotometrically or by polarography directly on the ion-exchange effluent. Alternatively it may be possible to extract the purified pesticide from the aqueous medium into an organic solvent when all the range of physical techniques are available for final determination.

The use of ion-exchange resins has proved to be particularly valuable in determining residues of the bipyridyl herbicides, which are highly polar compounds insoluble in the common organic solvents. Application of the technique need not be restricted, however, to this type of compound. Thus resins may be used for concentrating substances from dilute solutions of plant extract which are much more weakly basic and which have limited solubility in water. A cation-exchange resin would almost certainly be of value in concentrating residues of the growth regulator cycocel (2-chloroethyl-trimethylammonium chloride) and possibly residues of the weakly basic triazine herbicides or the insecticide diazinon from food crops prior to analysis.

An obvious example, which has so far not been exploited, is the use of anion-exchange resins for the determination of the acidic herbicides such as MCPA, 2,4-D, the halogenated aliphatic acids, and Tordon. THORNBURG (1963) states that anion-exchange resins can sometimes be used to adsorb acidic herbicides from water extracts of plant tissue but he had found it difficult to obtain quantitative recoveries of materials like 2,4-D from anion resins. Hydrolysis products of the organophosphorus insecticides are readily adsorbed and eluted from anion-exchange resins and this technique has been of value in a study of their metabolism in plants and animals.

Acknowledgements

Some of the subject matter discussed is from the unpublished results of my colleagues Dr. J. W. BULLOCK, Dr. P. SLADE, Dr. J. B. TURNER, and Dr. S. H. YUEN.

Summary

The uses which have been made of ion-exchange resins for concentrating and purifying pesticide residues are reviewed. Some of the factors involved in selecting the appropriate resin for any particular application are also discussed. Although the technique has proved particularly valuable in analysing residues of the bipyridyl herbicides it need not be limited to strongly polar compounds or to compounds which are very soluble in water.

Résumé *

Les usages qui ont été faits des résines échangeuses d'ions pour concentrer et purifier les résidus de pesticides sont passés en revue. Certains des facteurs impliqués dans le choix de la résine appropriée à une application particulière sont aussi discutés. Bien que la technique se soit montrée spécialement valable dans l'analyse des résidus d'herbicides dérivés du bipyridyl, elle ne doit pas être limitée aux composés fortement polaires ni à ceux qui sont très solubles dans l'eau.

Zusammenfassung **

Es wird eine Übersicht gegeben über die Anwendung von Ionenaustauscher-Harzen bei der Konzentrierung und Reinigung von Pflanzenschutz-

* Traduit par S. DORMAL-VAN DEN BRUEL.
** Übersetzt von H. F. LINSKENS.

mittel-Rückständen. Einige Faktoren, die für die Auswahl geeigneter Austauscher bei speziellen Problemen maßgebend sind, werden diskutiert. Obgleich diese Technik sich bei der Analyse von Bipyridyl-Herbiziden als besonders wertvoll erwiesen hat, braucht sie nicht auf streng polare oder sehr gut wasserlösliche Verbindungen beschränkt zu bleiben.

References

Adams, B. A., and E. L. Holmes: Adsorptive properties of synthetic resins. J. Soc. Chem. Ind. (London) **54**, IT (1935).

Calderbank, A., C. B. Morgan, and S. H. Yuen: Determination of diquat residues in potato tubers. Analyst **86**, 569 (1961 a).

—, E. C. Edgar, and J. A. Silk: Triazinyl thiophosphate esters with selective aphidical properties. Chem. Ind. 630 (1961 b).

—, and J. B. Turner: An ion-exchange method for determining menazon residues in food crops. Analyst **87**, 273 (1962 a).

—, and S. H. Crowdy: Bipyridylium herbicides. Ann. Rept. Progress Applied Chem. XLVII, 536 (1962 b).

—, and S. H. Yuen: An ion-exchange method for determining paraquat residues in food crops. Analyst **90**, 99 (1965).

— Hydrolysis products of menazon. J. Chem. Soc. [c], 56 (1966).

Calmon, C., and T. R. E. Kressman: Ion exchangers in organic and biochemistry. New York: Interscience 1957.

Dauterman, W. C., J. E. Casida, J. B. Knaak, and T. Kowalczyk: Bovine metabolism of organophosphorus insecticides. Metabolism and residues associated with oral administration of dimethoate to rats and three lactating cows. J. Agr. Food Chem. **7**, 188 (1959).

—, G. B. Viado, J. E. Casida, and R. D. O'Brien: Persistence of dimethoate and metabolites following foliar application to plants. J. Agr. Food Chem. **8**, 115 (1960).

Davies, H. M.: Recent advances in polarography. Royal Inst. Chem. J. **88**, 104 (1964).

Fox, H. M.: Substituted carbamoylmethyl bipyridyls PP 407 and PP 745. Proc. 7th Brit. Weed Control Conf. **1**, 29 (1964 a).

—, and C. R. Beech: Bipyridylium herbicides: Field trials with PP 407 and PP 745 as selective herbicides for weed control in cereals. Proc. 7th Brit. Weed Control Conf. **1**, 108 (1964 b).

Herrett, R. A., and A. J. Linck: The metabolism of 3-aminotriazole by Canada thistle and field bindweed and possible relation to its herbicidal action. Physiologia Plantarum **14**, 767 (1961).

—, and W. A. Bagley: The metabolism of 3-aminotriazole by Canada thistle. J. Agr. Food Chem. **12**, 17 (1964).

Hodgson, E., and J. E. Casida: Mammalian enzymes involved in the degradation of 2,2-dichlorovinyl dimethyl phosphate. J. Agr. Food Chem. **10**, 208 (1962).

Krueger, H. R., J. E. Casida, and R. P. Niedermeier: Bovine metabolism of organophosphorus insecticides. Metabolism and residues associated with dermal application of Co-Ral to rats, a goat, and a cow. J. Agr. Food Chem. **7**, 182 (1959).

Marco, G. J., and E. G. Jaworski: Metabolism of O-phenyl-O'-(4-nitrophenyl) methylphosphonothionate (Colep) in plants and animals. J. Agr. Food. Chem. **12**, 305 (1964).

Ospenson, J. N., and D. E. Pack: Paraquat-recovery from soil and clay samples. *California Chemical Co.*, Ortho Division, Richmond, Calif. 1964.

Pack, D. E.: Analysis of paraquat residues, Method RM-8. *California Chemical Co.*, Ortho Division, Richmond, Calif. 1964.

Plapp, F. W., and J. E. Casida: Ion-exchange chromatography for hydrolysis products of organophosphorus insecticides. Anal. Chem. **30**, 1622 (1958 a).

PLAPP, F. W., and J. E. CASIDA: Bovine metabolism of organophosphorus insecticides. Metabolic fate of O,O-dimethyl O-(2,4,5-trichlorophenyl) phosphorothioate in rats and a cow. J. Agr. Food. Chem. **6**, 662 (1958 b).

RACUSEN, D.: The metabolism and translocation of 3-aminotriazole in plants. Arch. Biochem. Biophys. **74**, 106 (1958).

SALMON, J. E., and D. K. HALE: Ion-exchange — a laboratory manual. London: Butterworths 1959.

SAMUELSON, O.: Ion-exchange separations in analytical chemistry. 2nd Ed. New York: Wiley 1963.

SLADE, P.: Photochemical degradation of paraquat. Nature **207**, 515 (1965 a).

— Fate of paraquat in plants. Submitted for publication (1965 b).

SPRINGETT, R. H.: The bipyridyl herbicides: their properties and use. Outlook on Agr. IV (5), 226 (1965).

STORHERR, R. W., and J. BURKE: Determination of 3 aminotriazole in crops. *U. S. Dept. of Health, Education and Welfare,* Food and Drug Administration, Washington D. C. 1960.

THOMAS, M. P., and H. J. ACKERMANN: Determination of beta-hydroxyethyl hydrazine in pineapples. J. Agr. Food Chem. **12**, 432 (1964).

THORNBURG, W. W.: Analytical methods for pesticides, plant growth regulators, and food additives. Ed. G. ZWEIG. Vol. I, p. 108. New York and London: Academic Press 1963.

Determination of organophosphorus insecticide residues using Cooke's emission spectroscopic detector

By

CARL A. BACHE and DONALD J. LISK*

With 8 figures

Contents

I. Introduction

Emission spectroscopy is an old technique for the analysis of metals and some nonmetallic elements. The sample to be analyzed is excited using a spark, direct current arc, flame, or other means. Photons are emitted when electrons in the excited species of the sample return to a lower energy state. This emitted quantized energy may then be measured photographically, oscilloscopically, or by automatic recording. The frequency of radiation can provide a high degree of specificity for characterizing the species and radiation intensity is used for quantitative analysis.

The radiation may appear as discrete lines or as molecular bands. Lines result from purely electronic transitions in excited atoms or atomic ions. Bands arise from excited molecules and, since two or more bonded nuclei result in vibrational and rotational as well as electronic changes, the bands appear as groups of lines located closer and closer together approaching the head of the band which represents complete ionization.

Although sensitive, the detection limits of emission methods in general have not in the past been adequate for analysis of pesticide residues. This difficulty was largely overcome when COOKE and co-workers (MCCORMACK *et al.* 1965) employed a microwave-powered argon discharge as an emission source with a high resolution spectrometer as part of a gas chromatography detection system. They reported sensitivities for hydrocarbons and organic compounds containing halogens, sulfur, nitrogen, and phosphorus to levels ranging from 10^{-9} to 10^{-16} g./sec. This equipment was found very effective for selective determination of organophosphorus insecticide residues (BACHE and LISK 1965). The details of operation and results are described here.

* Pesticide Residue Laboratory, Cornell University, Ithaca, New York.

II. Apparatus

Figure 1 shows a schematic diagram of the chromatograph and detection system and in Figure 2 a photograph of the equipment is presented. A Research Specialties Model 601-1 column oven and proportional temperature controller were used. The large hinged metal oven cover was replaced with a piece of 0.5-inch transite.

The column was borosilicate glass, U-shaped, five mm. in inside diameter (i. d.), and two feet long. The packing was five percent of the ethyl acetate-soluble fraction of Dow Corning high vacuum silicone grease or S. E. 30 silicone gum rubber on 80—100 mesh acid-washed Chromosorb *W*. Sample injection was made directly on the column through a silicone rubber septum. The carrier gas was Linde argon controlled by a Foxboro Model B102RP restrictor. Flow rates from 20 to 115 cc./minute were used for analysis of samples.

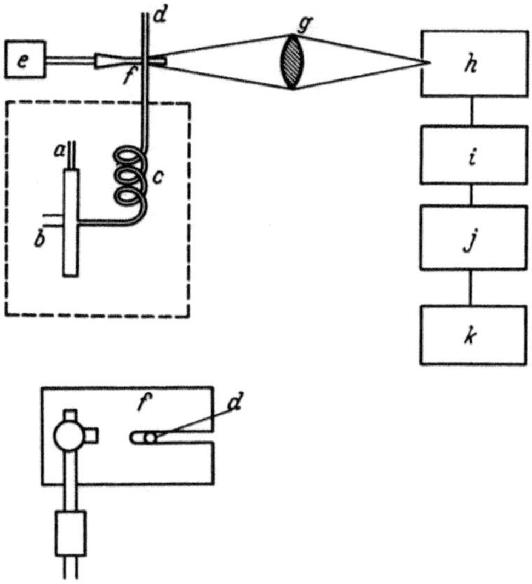

Fig. 1. Schematic diagram of detector: a) Carrier gas inlet, b) injection port, c) column, d) quartz discharge tube, e) microwave generator, f) microwave cavity, g) quartz lens, h) monochrometer, i) photomultiplier, j) amplifier, and k) recorder

The connection between the column outlet and sample excitation (discharge) tube was made in the column oven using speptums and a short length of Teflon tubing. The discharge tube consisted of a ten-inch length of 0.8 mm. bore, three mm. outside diameter (o. d.), transparent quartz capillary tubing (Thermal American Fused Quartz Co., Montville, N. J.). The precision (as regards the specified bore and wall thickness) of these tubes as manufactured may vary up to ±20 percent. This tube extended vertically eight inches above the column oven through the center of a five-inch length of 20-mm. i. d. borosilicate tubing. This chimney was open to the oven at the bottom and the discharge tube was held in the center of it with a fabricated asbestos plug at its top. About three inches of the discharge tube protruded above the chimney. The chimney was heated outside with heating tape to prevent condensation of sample components before reaching the discharge.

A Raytheon Model PGM-10 85-watt microwave power generator with tapered matching section was used. The generator's magnetron produced microwaves at a frequency of 2450±25 megacycles. Line voltage was stabilized through a Sola CVS transformer. A filter system was installed in the generator to reduce ripple to below one percent. The tapered matching section was centered around the sample tube 0.5 inch above the chimney. The discharge induced in the tube heated it sufficiently down into the chimney to further prevent condensation. A 0.5-meter Jarrell Ash Ebert scanning spectrometer with a 30,000 grooves-per-inch grating, dual adjustable straight slits, and a selected 1P28 photomultiplier tube was used. The discharge was focused on the slit, and perpendicularly to it, with a three-inch diameter, four-inch focal length, fused quartz, plano convex lens (Esco

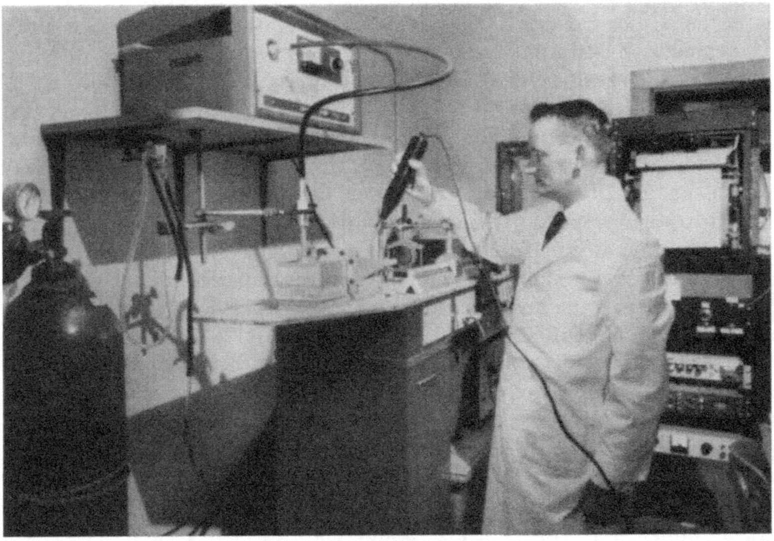

Fig. 2. Gas chromatograph and detection system

Products, Oak Ridge, N. J.). A Fluke Model 409A power supply was used at 680 volts for the photomultiplier tube. A Philbrick UPA-2 amplifier with a R100B power supply was used for signal amplification. The recorder was a Leeds and Northrup Speedomax G zero-to-one millivolt, with one-second response, and with ten-inch chart paper running 15 inches per hour.

III. Procedure

For analysis, up to five microliters of the sample was injected with the discharge off. After allowing about two minutes for the solvent to pass through, the microwave generator was turned on and the power was adjusted to about 80 percent. The discharge was ignited by directing a spark down into the discharge tube with a Tesla coil. The power was then readjusted to the optimum setting. The chromatogram was then recorded at a predetermined wavelength. The interval between sample injection and ignition of the discharge prevented the solvent from extinguishing the discharge and eliminated deposition of carbon in the quartz tube. By following this procedure the tube was used daily for three months with no visible carbon deposits. Complete fragmentation of organic compounds in the discharge permitted repetitive sample injection and rapid analysis. Solvents such as diethyl ether, acetone, chloroform, or hexane were all suitable for injection of samples.

A spectrum of malathion [*S*-(1,2-dicarbethoxyethyl)O,O-dimethyl dithiophosphate] insecticide was produced (McCORMACK *et al.* 1965) using similar equipment by vaporization of the compound directly into the discharge and recording the spectrum. Four emission lines appeared at 2553.98, 2535.65, 2553.31, and 2554.93 Angstroms (Å) corresponding to atomic phosphorus lines (AHRENS and TAYLOR 1961). The 2535.65 and 2553.31 Å lines were most intense. The former (2535.65 Å) was the stronger of the two and was used exclusively for analysis of organophosphorus insecticides in this investigation unless otherwise indicated. This line at 2535.65 Å is also reported (AHRENS and TAYLOR 1961) as the most usable sensitive line for emission analysis of phosphorus. The precise wavelength setting for optimum sensitivity was determined by first locating the approximate 2535.65 Å region using the 2536.52 Å mercury line. The same amount of a given compound was then repeatedly injected while varying the wavelength about 0.2 Å each time in the 2535.65 Å region using a slit width of about 25 microns. The setting showing the greatest peak height was used. Proper focus may also be checked for other wavelength regions by using a strong argon line such as 5650.70 Å.

The effect of ambient temperature changes and normal laboratory vibrations on the spectrometer grating made periodic refocusing essential. A change of four degrees C. in ambient temperature, for instance, can markedly change the focus and therefore the response. Permanent mounting of the discharge tube, lens, and spectrometer on an optical bench and operation in a constant temperature room would be very advantageous. The presence of small amounts of nitrogen in all commercial argon is a cause for the usual presence of C-N band spectra. The 2535.65 Å region was free of other strong interfering lines such as certain of those present in C-C or C-N excitation bands. A wide slit opening (75 microns) could therefore be used to increase sensitivity.

IV. Results and Discussion

In Fig. 3 is illustrated a gas chromatogram of ethion (O,O,O′,O′-tetraethyl *S,S′*-methylene bisphosphorodithioate) and the reagent blank. Fig. 4 shows standard curves for diazinon [O,O-diethyl O-(2-isopropyl-4-methyl-6-pyrimidinyl)phosphorothioate], and parathion [O,O-diethyl O-(p-nitrophenyl)thiophosphate] insecticides. The standard curves were very reproducible.

The response of the detector to several organophosphorus insecticides was determined at a signal to noise ratio of three-to-one. These data are presented in Table I. Those compounds containing the P-S (thiophosphorus) group show about the same response as those containing P-O (Ciodrin, Dibrom, Phosdrin, and Phosphamidon).

The type of bonding and the groups attached to an atom may affect its emission response (McCORMACK 1963). These effects are most pronounced when dealing with band spectra. The power settings listed in Table I were those found to be optimum for maximum response. Theoretically, when dealing with atomic spectra, the more power that can be transferred from

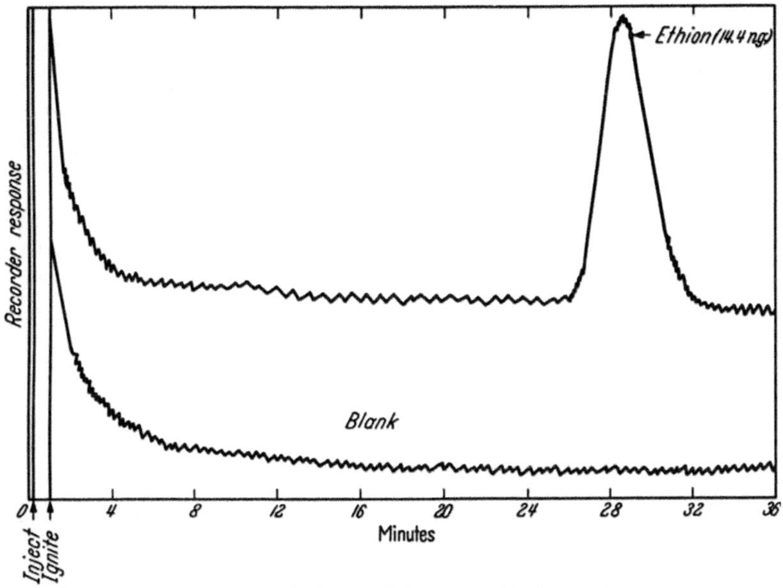

Fig. 3. Gas chromatograms of ethion and the reagent blank; "ng." = nanograms

Table I. *Detector response to organophosphorus insecticides*

Compound	Microwave power, %	Sensitivity in g. phosphorus/sec. (2535.65 Å emission)
Ciodrin	50	9.2×10^{-12}
Diazinon	90	6.6×10^{-12}
Dibrom	50	1.6×10^{-11}
Dimethoate	50	2.3×10^{-12}
Disyston	26	5.5×10^{-12}
Disyston sulfoxide	50	1.9×10^{-11}
Disyston sulfone	50	2.2×10^{-11}
Ethion	90	2.0×10^{-12}
Malathion	50	2.5×10^{-12}
Methyl parathion	24	8.5×10^{-12}
Parathion	90	1.4×10^{-12}
Phosdrin	50	4.6×10^{-12}
Phosphamidon	50	3.3×10^{-11}
Ronnel	21	1.5×10^{-12}
Systox	20	2.0×10^{-11}
Thimet	22	6.5×10^{-12}
Thimet oxygen analog	50	7.6×10^{-12}
Trithion	37	3.6×10^{-12}

the generator to the discharge tube, the greater the number of atoms excited and the stronger the emission. An optimum power setting for maximum response was found for the organophosphorus insecticides which depended upon other operating parameters such as column temperature and flow rate. This relationship has been suggested as a simple measure of the impedance match between the microwave cavity and discharge tube and its contents (EASTMAN 1964). The effect of column temperature on maximum response may have been largely related to the temperature at which partial compound decomposition on the column occurred. Discharge tube geometry is also important. Thin walls and small bore contribute to greater response, since less power is absorbed. In the discharge, fragmentation, excitation, and emission result from bombardement by high energy electrons (those provided by the Tesla coil), but the process is probably very complex (MCCORMACK *et al.* 1965).

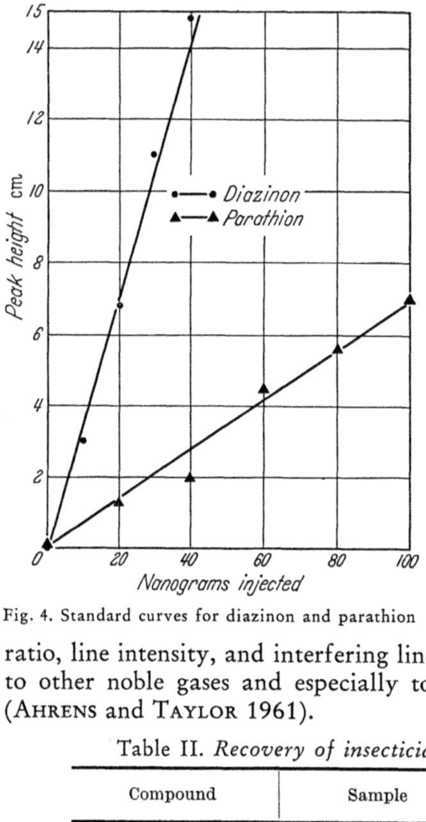

Fig. 4. Standard curves for diazinon and parathion

The response of the emission spectroscopic detector is in the same range as that of the sodium thermionic detector (GIUFFREDA 1964). The high signal-to-noise ratio attainable with the microwave-powered argon discharge at this wavelength setting allowed considerable signal amplification. The resultant sensitivity was considerably greater than that possible with other excitation sources such as flame or arc. The advantages as regards signal-to-noise ratio, line intensity, and interfering lines of excitation in argon as compared to other noble gases and especially to air have been discussed previously (AHRENS and TAYLOR 1961).

Table II. *Recovery of insecticides from agricultural samples*

Compound	Sample	Added p.p.m.	Recovery %
Diazinon	grapes	0.30	79
Dimethoate	alfalfa	0.06	106
		0.19	88
	timothy	0.20	91
	lettuce	0.03	113
	milk	0.20	115
	cow urine	0.20	98
Disyston	potatoes	0.18	81
	grapes	0.40	72
Ethion	soil	0.20	73
Parathion	bees	0.60	83
	halibut	0.20	105
	lettuce	0.20	82
Ronnel	whole chicken	0.25	71
	eggs	0.25	90

Figures 5, 6, and 7 show typical chromatograms of organophosphorus insecticides having been added and recovered from agricultural samples and the respective control samples. Table II lists the recovery of insecticides added to samples prior to extraction. The extraction and isolation procedures used for these samples have been described (BACHE and LISK 1965).

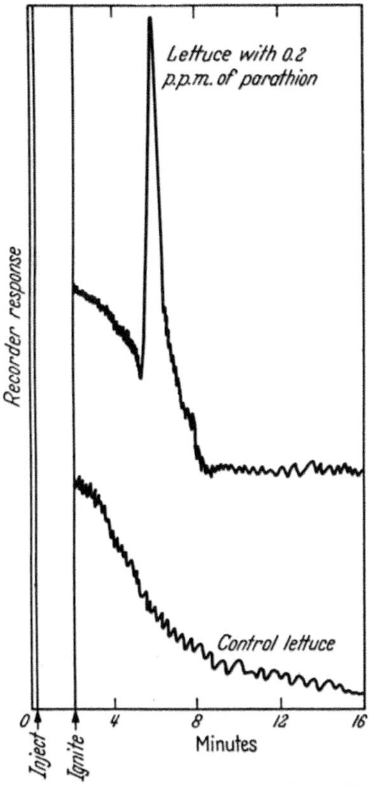

Fig. 5. Gas chromatograms of parathion added to lettuce and of control lettuce

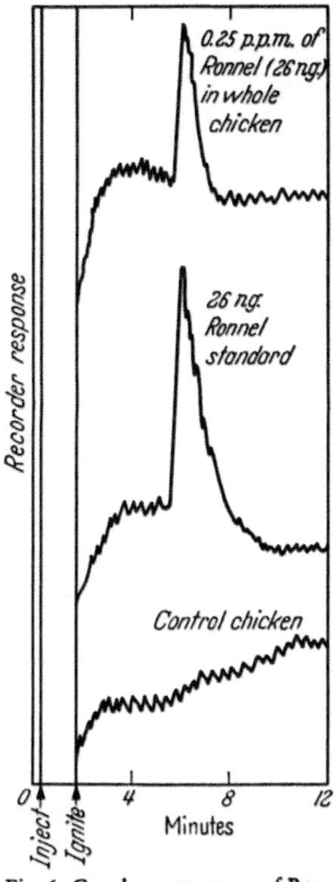

Fig. 6. Gas chromatograms of Ronnel added to whole chicken, standard Ronnel, and control chicken

It should be noted that although the response of hydrocarbons observed (TONG *et al.* 1965) at 3883 Å and 5164 Å were very high (2×10^{-16} and 1×10^{-14} g./sec., respectively), injection of green solvent extracts of lettuce into the S. E. 30 column at these wavelengths resulted in chromatograms which were remarkably free of major peaks. This implicates the column as an important contributor to selectivity, presumably by its irreversible adsorption of many extraneous plant substances as reported by many investigators.

The detection system is effective for the accurate, sensitive, and selective determination of many organophosphorus insecticides in food samples. The attainment of a sensitivity of 0.01 p.p.m. for many organophosphorus compounds as residues in crops is not unrealistic and will depend largely on the rigor of the accompanying isolation procedure. Increases in sensitivity of the detection system could logically be expected from use of a photomultiplier tube having a higher signal-to-noise ratio.

The use of a high energy excitation source is advantageous when attempting to induce atomic emission in nonmetals such as phosphorus or sulfur with their high ionization potentials. The selectivity of the detection system can

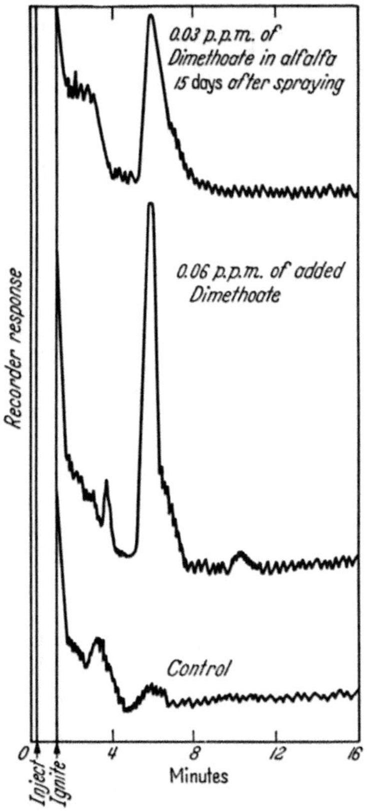

Fig. 7. Gas chromatograms of Dimethoate in field-treated alfalfa, the insecticide recovered (106 percent) from alfalfa, and control alfalfa

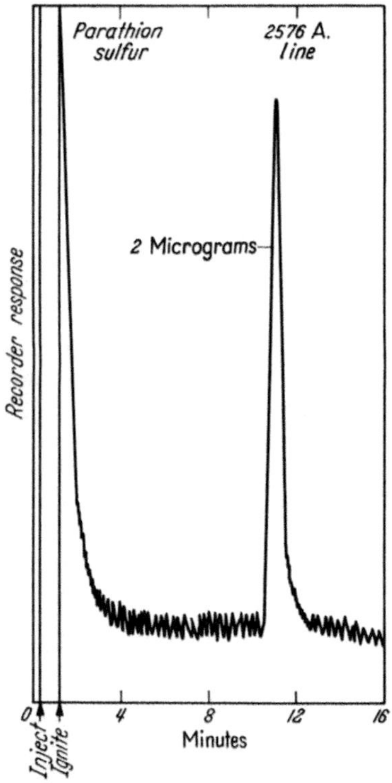

Fig. 8. Gas chromatogram of 2 µg. of parathion resulting from sulfur emission at 2576 Å

be controlled, in part, from each of the three operating parameters: retention time, wavelength, and power setting. Owing to the wide dynamic range (McCORMACK *et al.* 1965), the deliberate use of less sensitive emission lines for other applications such as formulation analysis may be appropriate. For example, the chromatogram resulting from injection of two micrograms of standard parathion and measurement of the 2576 Å C-S band emission is illustrated in Figure 8. Emission response at 2576 Å is about 0.01 the sensitivity of the 2535.6 Å phosphorus emission. Depending on the pesticide concentration and line sensitivity, the identification of residues or metabolites using different analytical wavelengths for the same molecule is therefore a further possibility.

The type of excitation used in this study is an example of a high frequency, electrodeless discharge. The magnetron is a generator of continuous high power pulses of microwave energy. Its life, like that of an amplifier tube, is unpredictable. Satisfactory operation for a year or longer under conditions of continuous daily use is not uncommon, however. The 2450-megacycle microwave frequency can be very damaging to the eyes (cataract formation), liver, and certain other internal organs through internal heating. This is especially true at higher wattages. A grounded metal shield with a small 16-mesh metal view screen conveniently positioned between the tapered section (but several inches from it so as not to interfere with microwave transmission) and the operator will offer sufficient protection.

For continuous use at high power settings (90 percent), a thick-walled (8.0-mm. bore, eight-mm. o.d.) quartz capillary discharge tube was necessary to provide adequate heat capacity and therefore prevent deterioration of the tube by the intense heat of the discharge. When operating at 20 percent power, the 0.8-mm. bore, three-mm. o.d. quartz tube showed no deterioration in continuous use. Further studies with organophosphorus and other classes of pesticide compounds will be made to determine the location of possible sensitive lines for other elements and the operating conditions for sensitive determination of their residues.

Summary

The microwave-powered argon emission detector of Cooke is sensitive and selective for organophosphorus insecticide residues. Measuring the intensity of the 2535.65 Å atomic phosphorus line during gas chromatography provides a sensitive and specific method for these compounds. Selectivity is determined by retention time, wavelength, and microwave power setting. The apparatus has been used for analysis of a wide variety of agricultural commodities to which insecticides were added from levels of 0.03 to 0.6 p.p.m. Its potential use at other wavelength settings is under investigation.

Résumé*

Le détecteur à spectroscopie d'émission de Cooke est sensible et spécifique pour les résidus d'insecticides organophosphorés. La mesure de l'intensité de la raie 2535,65 Å de l'atome de phosphore après la chromatographie gazeuse donne une méthode sensible et spécifique pour ces composés. La sélectivité est obtenue par la valeur du temps de rétention, la longueur d'onde et le réglage de la puissance de la micro-onde. L'appareil a été utilisé pour l'analyse d'une grande variété de produits agricoles auxquels des insecticides avaient été ajoutés à des taux variables de 0,03 à 0,6 ppm. Ses possibilités d'utilisation à d'autres longueurs d'ondes sont à l'étude.

* Traduit par R. Mestres.

Zusammenfassung *

Der Mikrowellen-Argon Emissionsdetektor von COOKE ist empfindlich und selektiv für Organophosphor-Insektizid-Rückstände. Die Messung der Intensität der 2535,65 Å Phosphorlinie bei der Gaschromatographie liefert eine empfindliche und spezifische Methode für diese Verbindungen. Die Selektivität wird bestimmt durch die Retentionszeit, Wellenlänge und Mikrowellen-Einstellung. Der Apparat wurde zur Analyse einer breiten Vielfalt von landwirtschaftlichen Produkten benutzt, denen Mengen von 0,03—0,6 ppm an Insektiziden zugegeben wurden. Seine mögliche Anwendung bei anderen Wellenlängen wird noch untersucht.

References

AHRENS, L. H., and S. R. TAYLOR: Spectrochemical analysis. Reading, Mass.: Addison-Wesley 1961.

BACHE, C. A., and D. J. LISK: Determination of organophosphorus insecticide residues using the emission spectrometric detector. Anal. Chem. **37**, 1477 (1965).

EASTMAN, L. F. (Elec. Eng. Dept., Cornell Univ., Ithaca, N. Y.): Personal communication 1964.

GIUFFREDA, L.: A flame ionization detector highly selective and sensitive to phosphorus — a sodium thermionic detector. J. Assoc. Official Agr. Chemists **47**, 293 (1964).

McCORMACK, A. J.: Emission detectors for gas chromatography. M. S. Thesis Cornell Univ., Ithaca, N. Y. 1963.

TONG, S. C., and W. D. COOKE: A sensitive selective gas chromatography detector based on emission spectroscopy of organic compounds. Anal. Chem. **37**, 1470 (1965).

* Übersetzt von F. BÄR.

Inhibition vs. excessive use of vitamin K and other nutrients for the control of carpet beetles

By

Roy J. Pence* and Manuel S. Viray*

With 10 figures

Contents

I. Introduction

In the broad sense, antimetabolites are substances that inhibit the utilization of metabolites, usually by antagonistic action. They are compounds which chemically resemble essential nutrients such as vitamins and amino acids and, when introduced into a biological system, prevent these food sources from performing their normal function. In effect, antimetabolites are "metabolic disruptors". Conversely, some metabolites by themselves are

* Department of Agricultural Sciences, University of California, Los Angeles.
The authors gratefully acknowledge the assistance of Lance Ralston in the statistical compilations herein contained.

capable of disrupting a system when administered in excess, and induce symptoms which are directly or indirectly related to hypervitaminosis. This paper explores both views as a new approach to pest control wherein non-pesticidal chemicals are used which, by their nature, minimize the problems of toxic residues.

II. Modification of CSMA standards to comply with metabolic inhibition

The writers have had to modify CSMA *(Chemical Specialties Manufacturing Association)* standards to some extent when investigating the control of fabric pests via metabolic inhibition. CSMA standards require that ten test insects be confined to treated cloth specimens for an exposure period of 14 days. As all tests are made with insecticide-treated test cloth, it is understandable that insect mortality is an important factor in evaluating proofing effectiveness. Mortality is correlated with excrement weight, yields, and fiber weight losses. When substituting non-pesticidal metabolic inhibitors for conventional insecticides, the ultimate criterion is lack of fiber weight loss rather than mortality. This does not imply that mortality will not occur, for in all cases it is inevitable — but not always within the 14 days allowed. Sample feeding by the larvae must of necessity discontinue early if fiber weight losses are to be held within tolerances and this alone indicates that the test insects are in a state of decline, not from pesticide poisoning but rather from metabolic disruption. Under these circumstances, 100 percent mortality occurs after the 14 days allowed under CSMA standards. For most of the tests discussed in this paper, 25 rather than ten larvae are used for each test. Body weight losses and gains of surviving insects must be taken into account which predict rather than demonstrate mortality. For these studies, the furniture carpet beetle *Anthrenus flavipes* Lec. and the black carpet beetle *Attagenus piceus* (Oliv.) were used as test insects.

III. Free larval feeding selection

A technique employing free larval feeding selection is used to determine the reversal effects of a nutrient over its inhibitor. This is accomplished by first treating test cloth with a given antimetabolite. A small amount of the "cancelled" nutrient is then added in the center of the treated swatch of

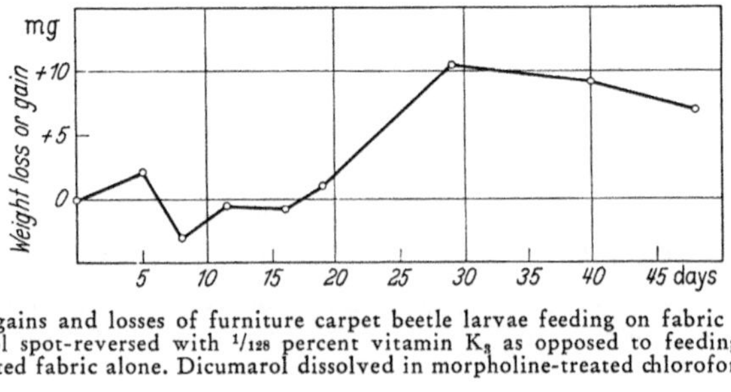

Fig. 1a. Weight gains and losses of furniture carpet beetle larvae feeding on fabric treated with one percent dicumarol spot-reversed with 1/128 percent vitamin K_3 as opposed to feeding on one percent dicumarol-treated fabric alone. Dicumarol dissolved in morpholine-treated chloroform

cloth and is designed to provide a "life saving" target area which the inhibited larvae must locate and feed upon if they are to survive ultimate death resulting from metabolic disruption. Selective feeding by the larvae

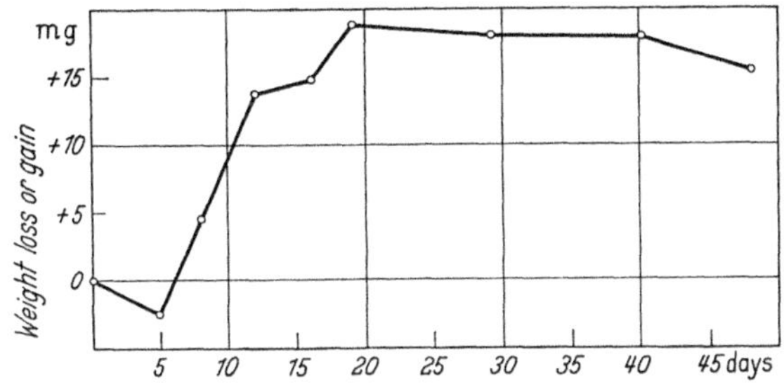

Fig. 1 b. Weight gains and losses of black carpet beetle larvae on fabric treated with one percent dicumarol spot-reversed with $^1/_{128}$ percent vitamin K_3 as opposed to feeding on one percent dicumarol-treated fabric alone. Dicumarol dissolved in morpholine-treated chloroform

clearly demonstrates the difference between an inhibited food and the target area overlay of the same nutrient being inhibited. Such a technique enables one to study not only the reversal effects of the overlay but also to observe the voluntary behavior of larvae which seek out their own replacement therapy (see Figs. 1 a and b).

a) Migratory differences between species of carpet beetle larvae

The furniture carpet beetle has a greater capacity for migration than the black carpet beetle, and is thus better suited for study under free larval feeding selection. When confined, furniture carpet beetle larvae are quick to detect "negative" food and move off to more suitable supplies. This accounts for less sampling and greater intake of counterbalancing nutrients once a target area is located. On the other hand, black carpet beetle larvae, with poorer migratory advantage but greater reserve of accumulated body food, will linger longer on inhibited food. However, some are incapable of ever finding a corrective target area and it is only accumulated food reserve that enables this species to live longer on inhibited food than the furniture carpet beetle, even though this reserve becomes less competitive during the passage of time. In spite of migratory differences between both species, the black carpet beetle is still the more difficult to kill when exposed to antimetabolites or, for that matter, many conventional insecticides when combined with a wool diet. The exception occurs when a state of imbalance is induced at the symbiont level such as by suppressing vitamin K synthesis with doses of vitamin A, further discussed in Section V. Whatever the approach, fiber weight loss becomes the principal criterion for evaluating proofing effectiveness, with body weight losses a convenient barometer for predicting ultimate mortality. *Rate* of antimetabolic intake from the surrounding area plus *time involved* for the larvae to discover and ingest enough of the replacement metabolite from the target area will govern the success of any deficiency correction. The longer any antagonist remains at the enzyme site the longer the period of imbalance which will in turn hasten decline. Mortality must therefore be correlated with the body weight losses of surviving insects.

b) Use of vitamin-free casein as a substrate in place of wool

Reluctance of the black carpet beetle immediately to accept a diet of wool when transferred from a rearing medium of ground dog food prompted trials with vitamin-free casein as a possible substitute for wool. In spite of its vitamin lack, vitamin-free casein is high in usable protein and serves temporarily as a substrate of immediate acceptability. On such a diet, black carpet beetle larvae will initially gain prior to ultimately losing weight due to vitamin deficiency. For this reason preliminary tests were made whereby 25 half-developed black carpet beetle larvae were placed on a one-percent vitamin A-impregnated vitamin-free casein diet as well as upon an identical diet having 1/4 percent vitamin K_1 added to compensate for the suspected symbiont-killing activity of vitamin A. In this instance, results were highly-

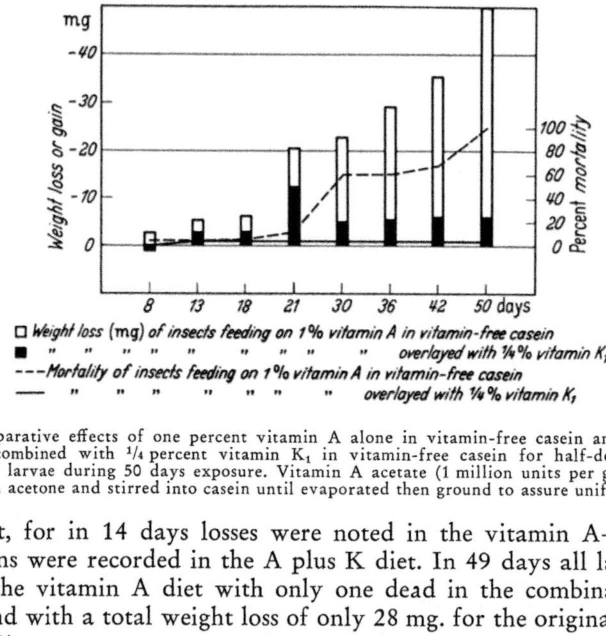

Fig. 2. Comparative effects of one percent vitamin A alone in vitamin-free casein and one percent vitamin A combined with 1/4 percent vitamin K_1 in vitamin-free casein for half-developed black carpet beetle larvae during 50 days exposure. Vitamin A acetate (1 million units per gram) was dissolved in acetone and stirred into casein until evaporated then ground to assure uniform blend

significant, for in 14 days losses were noted in the vitamin A-fed larvae while gains were recorded in the A plus K diet. In 49 days all larvae were dead in the vitamin A diet with only one dead in the combination of A and K, and with a total weight loss of only 28 mg. for the original 25 larvae (see Fig. 2).

c) Preliminary studies with other fat-soluble vitamins

The role of vitamin K and its inhibition in insects is in need of additional study. In this connection more work is required to better understand the importance of micro-nutrients made available through symbiotic association. Of the fat-soluble vitamins we have learned that vitamins K and A are directly and indirectly important to two species of fabric-feeding insects and, as of this writing, are finding vitamin E, with its structural and physiological relationship to vitamin K, to be of importance also. Preliminary work in inhibiting the tocopherols indicates some subsequent impairment to the reproductive system of the furniture carpet beetle. Vitamin D is also under investigation and has been shown to be an effective growth promoter when incorporated into the fabric insect's diet. In view of these findings work is being carried on in an effort better to understand the importance for insects of fat-soluble vitamins other than those which are sustained on a diet of wool. The objectives are to control insects with non-hazardous substances, which will lessen the problems encountered in the use of persistent poisons which too often leave residues of a harmful nature.

IV. The obscure role of vitamin K in insect biochemistry

Hitherto vitamin K has not been considered important as an essential nutrient for most insects. Of the fat-soluble vitamins (A, D, E, and K), no requirement has been demonstrated except in the case of *Schistocera,* which was shown to need β-carotene in the diet both for pigmentation and growth (GILMOUR 1961).

Indirect reference to a possible usability of vitamin K has been made by a number of workers, particularly with reference to symbiotic association within the gut, where it is assumed traces of vitamin K may be synthesized. GREGOIRE (1958) came close to defining the role of vitamin K in his studies with blood coagulation in arthropods. In this work he showed that 420 species of insects demonstrated varying degrees of hemolymph coagulation and that a possible relationship exists between the pattern of coagulation and the phylogeny of insects. Attention was given to the nature of hemocytes involved and the nature of plasmic reaction. Earlier and perhaps more significant work by GREGOIRE (1953) involved the reaction of insect hemolymph to certain coagulation inhibitors of vertebrate blood wherein 33 substances were investigated *in vitro* on the hemolymph of different species, chiefly Orthoptera. These studies revealed that anticoagulants were found to be effective when mixed with extracted hemolymph and, while several anticoagulants of vertebrate blood were not inhibitors when used at normal strength, they would inhibit or in some cases increase coagulation when used at greater concentrations. However, GREGOIRE did not report on investigation *in vivo.*

a) Outline of study

The importance of vitamin K to fabric-feeding insects is at present being determined by (1) examination of dietary requirements both primitive and current, (2) investigation of substrate competition by means of experimental diagnoses with vitamin K inhibitors, (3) investigation of the biosynthesis of vitamin K via symbiotic associations, and (4) the reversal of vitamin K inhibition by vitamin K replacement. Results of many aspects of this outline are here presented.

1. Historic view of vitamin K importance to fabric-feeding insects. — From an historical viewpoint, one has to examine in retrospect the time when fabric insects were dependent solely upon basic scavenger instincts prior to the advent of man's discovery of weaving wool into a useful fabric which was later to become a convenient and abundant supply of food for these insects. Originally "wool" was made available to scavenger insects from the carcasses of fallen animals that would putrefy, leaving only the enriched pelage for their food supply. Even today it is not uncommon to find dermestid infestations thriving on the carcasses of dead animals which are completely consumed to the skeletal remains after complete destruction of the fur. Since vitamin K, and particularly vitamin K_2, occurs in many bacteria, most putrefied animal matter contains considerable quantities of this natural substance (HARPER 1959). It is conjectured that, in the past, vitamin K might have been ingested in appreciable amounts by the scavenging insects. The present-day lack of vitamin K in free amounts on processed woolens does not imply that these same insects do not continue to have a need for this metabolite. In the vertebrate animal it is known that a dietary deficiency of vitamin K is unlikely since the vitamin is fairly well distributed in foods of all descriptions and that, moreover, the intestinal microorganisms synthesize considerable vitamin K in the gut. A deficiency condition will, however, result whenever there is a bilary tract obstruction or a defect in fat absorption (HARPER 1959).

2. Present views based on dietetic studies. — The question arises as to which usable traces of vitamin K are to be found in the omnipresent contaminants that are associated with wool fibers. Molds, yeasts, and fungi contain traces, while quantities are present in many bacteria. Microorganisms are always present in varying degrees with the common contaminants of wool (PENCE 1959), and it is conceivable therefore that vitamin K is made available to fabric insects to some extent at the ingestion level as well as through the synthesis of vitamin K by the microflora of the gut.

Dietetic studies herein described clearly show that these insects have a need for vitamin K at both the ingestion and biosynthetic levels and that inhibition of this essential vitamin will cause their decline. It is also shown that symptoms of deficiency can be reversed by means of vitamin K replacement. The precise function of vitamin K within the insect system is not clearly understood; however, it is theorized that its function is probably twofold: (1) it may act in a manner comparable to that already recognized in mammalian physiology, and (2) the plausible explanation of GILMOUR (1961) who states that vitamin K may possibly play a role in insect aerobic energy production in that a quinone oxidation-reduction has been suggested as an intermediate, acting between the substrate and the cytochrome system.

b) Progess with present concepts

The best-known function of vitamin K in mammalian physiology is to catalyze the synthesis of prothrombin by the liver. In the absence of this vitamin, a hypoprothrombinemia occurs in which blood-clotting time may be greatly prolonged. Uncontrollable hemorrhage is a symptom of vitamin K deficiency. On this basis, it was thought that a possible anti-vitamin K-anticoagulation syndrome might be induced in fabric-feeding insects, a theory which has now been partially supported by experimental evidence. This information, abstracted from accounts of medical practices based on physiological findings, deals with the processes of anticoagulation when produced by the oral administration of dicumarol to warm-blooded animals.

1. Role of dicumarol as a mammalian anticoagulant and its effects on fabric-feeding insects. — Dicumarol is related chemically to the naphthoquinone derivatives with vitamin K activity and, because of the chemical resemblance, dicumarol acts as anti-vitamin K by the method of substrate competition. It is thought that in the liver dicumarol replaces vitamin K at the latter's normal site of action and thus prevents the vitamin from carrying out its normal physiological function. As dicumarol cannot be utilized by the liver in the synthesis of prothrombin, a syndrome of vitamin K deficiency develops, i. e., the plasma prothrombin level falls and blood coagulation is depressed (Wright 1956). The anticoagulant principles of dicumarol have been employed as the active agent in the formulation of certain rodenticides. These include Warfarin [3-(α-acetonylbenzyl)-4-hydroxy-coumarin], and Pival (2-pivalyl-1,3,4-hydroxycoumarindanedion). Both compounds are effective anticoagulants in warm-blooded animals.

In an effort to determine if any of these compounds would be effective on the hemolymph of insects, experiments were conducted using the larvae of both the furniture and the black carpet beetles as test animals with the supposition that there may exist a possible vitamin K requirement by these species. Dicumarol demonstrates some biological activity when ingested at one percent in the wool diet, whereas Warfarin is negative. The performance of Warfarin is to be expected, inasmuch as rat baits containing this product are often infested with thriving populations of the varied carpet beetle which feed upon the cereal content. The toxicity of Pival eliminates this compound from further tests, as to date no means of differentiating between mortality induced by anticoagulation or by acute poisoning have been found. Closer attention was given to dicumarol to determine whether or not its effectiveness could be attributed to actual anticoagulation or whether some other factor was responsible. According to Best and Taylor (1954 and 1961), in reference to vertebrate physiology, dicumarol does not act *in vitro* as an anticoagulant, nor in the body except after it has been partially metabolized. These authors also present strong evidence that in its metabolism, salicylic acid is produced and is the active agent; *in vitro* experiments have demonstrated the conversion of dicumarol quantitatively to salicylic acid. It has long been known that the administration of salicylates, e. g., acetylsalicylic acid (aspirin) and sodium salicylate, is often followed by spontaneous bleeding which has been shown to be associated with hypoprothrombinemia. Salicylates also cause a diminution in the number of platelets and the anticoagulant action of salicylates is antagonized by vitamin K_1.

2. Salicylic acid. — On the strength of this medical clinical information, it was decided to try salicylic acid as a substitute for dicumarol in the diet of carpet beetle larvae. If indeed salicylic acid is responsible for the anticoagulation of blood in the vertebrate animal, then it is logical to assume it might conceivably act in a related manner on insect blood, for the hemolymph of insects superficially resembles the blood of vertebrates, both being an aqueous plasma containing various types of hemocytes with differences occurring mostly in its manner of circulation and dependence upon free amino acids and dietetic relationships. Salicylic acid employed at the same dosage and in the same method as dicumarol was found to be

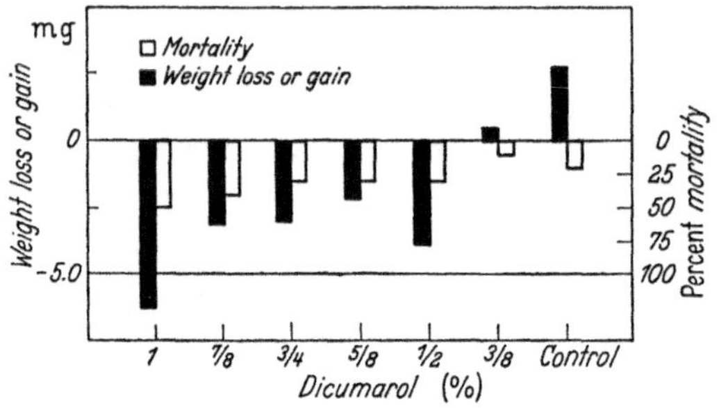

Fig. 3a. Effects of various concentrations of dicumarol in fabric on half-developed furniture carpet beetle larvae after 26 days exposure. Dicumarol dissolved in chloroform which was morpholine-alkalinated and surfactant-treated to assure, adequate distribution throughout test fabric

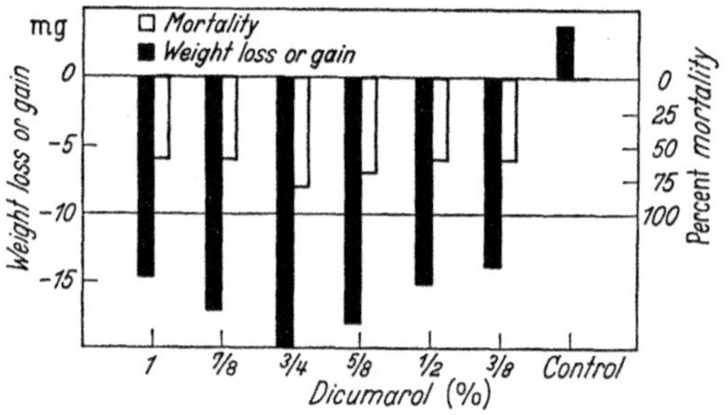

Fig. 3b. Effects of various concentrations of dicumarol in fabric on half-developed black carpet beetle larvae after 24 days exposure. Dicumarol dissolved in chloroform which was morpholine-alkalinated and surfactant-treated to assure adequate distribution throughout test fabric

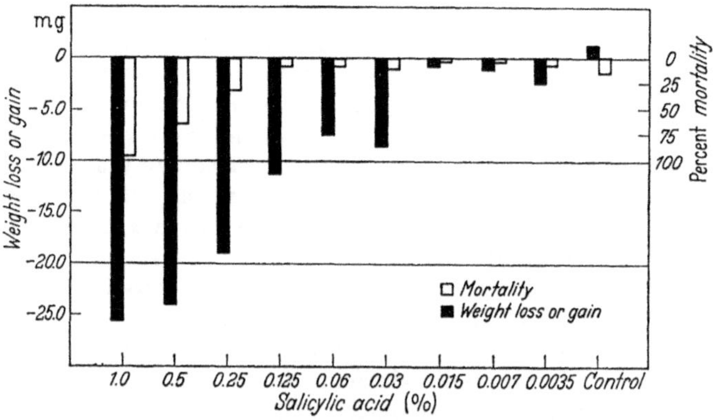

Fig. 4. Effects of various concentrations of salicylic acid on half-developed furniture carpet beetle larvae after 28 days exposure

approximately three times more effective. Figures 3 a and b, and 4 show the comparative effectiveness of dicumarol and salicylic acid.

Previous investigation by BEARD (1950) showed that dicumarol *in vivo* does not interfere with the clotting mechanism of hemolymph when this substance is injected directly into the insect system. However, administration of dicumarol via the oral route was not included in these studies and therefore the suspected anticoagulant salicylic acid by-product made available through partial metabolism of dicumarol remained unreported.

No implication is intended through the findings reported herein that vitamin K inhibition in carpet beetle larvae interferes with the clotting mechanism of hemolymph. Rather it is suggested that a syndrome may be involved between anticoagulation and interference with oxidative phosphorylation with speculation as to mechanisms involved remaining proportionate only to the current information at hand. Significant reversal of vitamin K deficiency by vitamin K replacement is apparent; however, the specific mode of action remains unexplained. The problem of obtaining sufficient amounts of hemolymph from minute larvae which in turn have suffered body weight losses due to vitamin K deficiency renders highly improbable the complex technique required for making actual platelet studies. Until such time as deleterious changes in hemolymph are precisely set forth, the phenomenon of anticoagulation remains only suspected.

3. Vitamin K balance, imbalance, and reversal. — It has been suggested that excessive amounts of vitamin K, if administered in the system, will antagonize the vitamin K that already exists. WOOLLEY (1952) discusses the toxicity of vitamin K when "relatively enormous" doses are given to overcome lethal doses of dicumarol. This in turn has been demonstrated by the writers while investigating vitamin K and its analogs when carpet beetle larvae are used as test animals. For these studies, three of the K vitamins were employed.

Vitamins K_1 and K_2 are both natural substances. Vitamin K_1 is derived from green leafy tissues of plants while K_2 is chiefly a product of putrefied animal matter. Vitamin K_3 (menadione) is synthetic and is found to be approximately three times as potent as either of the natural vitamins. The literature states that K_1 and K_2 are of about equal potency; however, in these studies K_1 was found to be dietetically more effective than K_2, but in some instances K_2 was found to demonstrate greater capacity for reversing an induced condition of vitamin K deficiency.

The potency of vitamin K_3 is such that, although only $^1/_{16}$ percent is incorporated into a test fabric diet, it will be sufficient to cause 100 percent mortality to 25 half-developed furniture carpet beetle larvae within 14 days' exposure. Significant ingestion of lethal amounts of K_3 is evidenced by discolored fecal matter found following feeding. However, the irritating properties of free menadione, plus its staining characteristics, may rule against the desirability of this compound as a practical "mothproofing" agent. Yet, interestingly enough, trace amounts of all three vitamins K, proportionate to their respective potencies, are found to be usable nutrients for this insect. This is shown by substantial weight gains of test insects when subjected to a broad spectrum of supplemental vitamin K under controlled dietetic studies (see Figs. 5 a, b, c, and d).

Assuming that carpet beetle larvae are capable of synthesizing a balanced amount of vitamin K, it is conceivable that the ingestion of excessive amounts might induce hypervitaminosis. This may explain why as little as

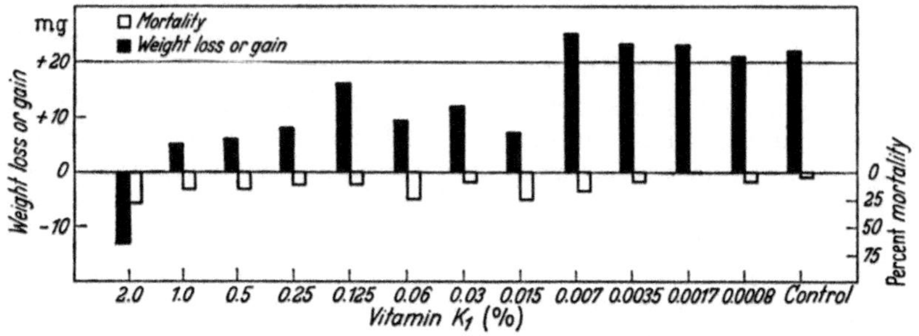

Fig. 5 a. Dietetic studies of half-developed furniture carpet beetle larvae during 56 days exposure to fabric treated with various concentrations of vitamin K_1. Vitamin K_1 expressed as actual derived from aqueous colloidal "solution" of Phytonadione (10 mg. per ml.) diluted in acetone for fabric penetrations

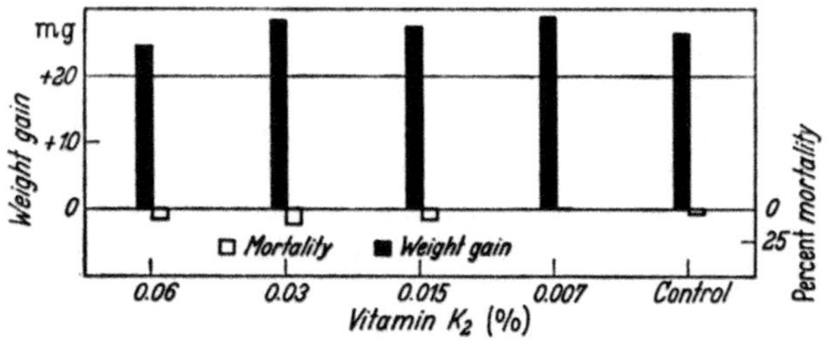

Fig. 5b. Dietetic studies of half-developed furniture carpet beetle larvae during 40 days exposure to fabric treated with various concentrations of vitamin K_2 in ethyl alcohol

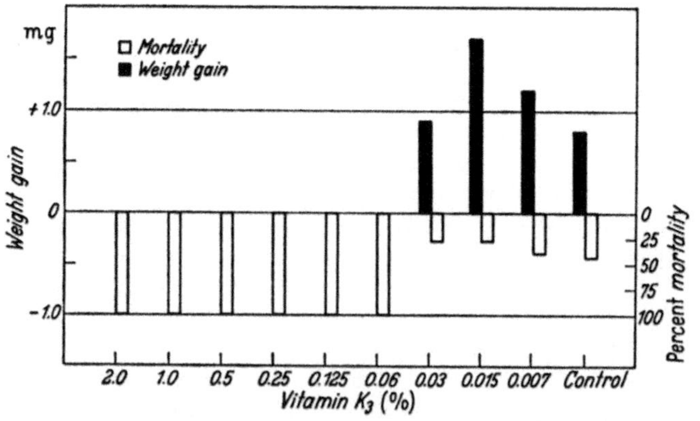

Fig. 5 c. Dietetic studies of first instar furniture carpet beetle larvae during 42 days exposure to fabric treated with various concentrations of vitamin K_3 (menadione)

$^1/_{16}$ percent menadione becomes fatal when added to a normal diet of half-developed carpet beetle larvae. Evidence of beneficial usability of trace amounts of the vitamins K added to the diet also appears to indicate that biosynthesis may not be quite sufficient to fulfill the insect's maximum capacity for complete utilization of this vitamin. It is speculated that this lack of biosynthetic efficiency might be attributed to a lag in complete

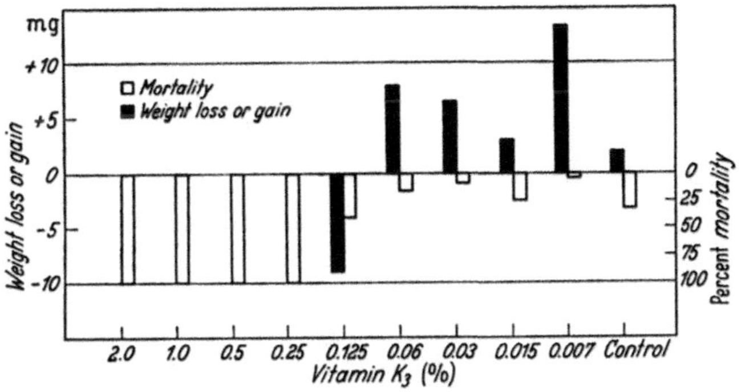

Fig. 5 d. Dietetic studies of half-developed furniture carpet beetle larvae during 28 days exposure to fabric treated with various concentrations of vitamin K_3 (menadione)

transition from the K-enriched primitive diet to the presently refined food supply which is lacking in available amounts of required vitamin K.

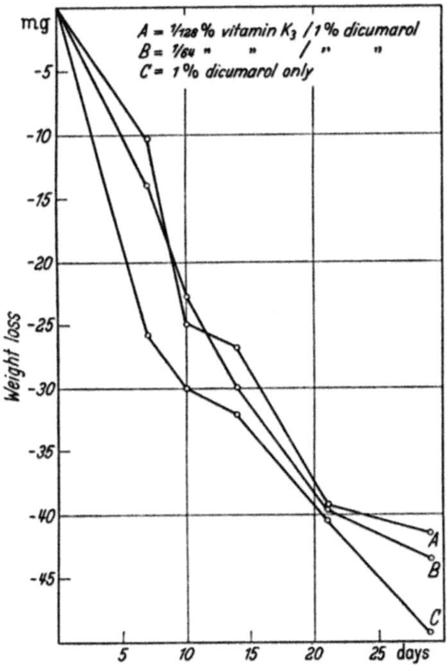

Fig. 6 a.

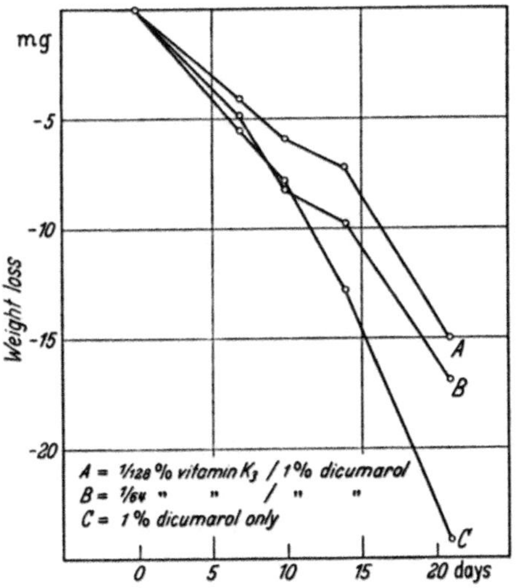

Fig. 6 b.

Fig. 6a. Partial reversal effects of vitamin K_3 (menadione) on half-developed black carpet beetle larvae when vitamin K_3 is added as counteractant to a one percent solution of dicumarol used to treat fabric

Fig. 6b. Partial reversal effects of vitamin K_3 (menadione) on half-developed furniture carpet beetle larvae when vitamin K_3 is added as counteractant to a one percent solution of dicumarol used to treat fabric

The need of both the black and the furniture carpet beetles for vitamin K is indicated by their significant decline when this important metabolite is inhibited and even more so by their recovery when vitamin K is restored to the diet (see Figs. 6 a and b, and 7).

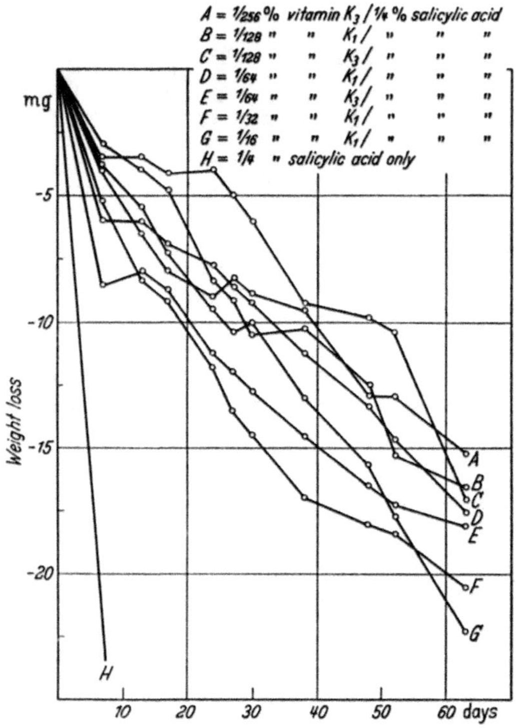

Fig. 7. Partial reversal effects of vitamins K_1, K_2, and K_3 on half-developed furniture carpet beetle larvae when K vitamins are added to one-half of fabric treated with $^1/_4$ percent salicylic acid. When portions of an inhibited swatch are overlaid with a counteracting metabolite, free larval feeding selection becomes evident. In this instance, all larvae migrated to the vitamin K-treated half of the swatch where feeding was concentrated

V. Symbiotic associations

BROOKS (1963) broadly discusses symbiosis and asymbiosis in arthropods and summarizes her work by stating that arthropods are veritable cultures of microorganisms, harboring rickettsia, bacteria, yeasts, and protozoa which are located in modifications of the intestines, Malpighian tubules, gonads, and in special cells in the fat body or organs in the hemocoele. Although numerous technical problems make analyses difficult, there is general agreement that the relationships are mutually beneficial, although not always equally so to each partner. BROOKS makes the general observation that symbionts are found in insects that eat a restricted diet deficient in certain essential nutrients, which suggests to workers in the field of arthropod symbiosis that symbiotic microorganisms are involved in synthesizing these deficiencies for the benefit of their hosts.

GUSTAFSSON *et al.* (1962), in a detailed study of the contribution by bacterial flora of the rat gut to the host's specific requirement for vitamin K, demonstrated for the first time that correction of an induced vitamin K deficiency could be realized by infecting the host with a single strain of bacterium. In this instance a strain of *Escherichia coli* and *Sarcina* sp. restored the prothrombin time to normal after 48 hours. It has been shown by chemical methods that both these organisms synthesize vitamin K.

VI. Eliminating symbionts with vitamin A

In order better to understand the role of symbiosis associated with black and furniture carpet beetle larvae, the extent of the importance of symbiotic microorganism synthesis of vitamin K was determined. In this attempt the objectives were to eliminate the symbionts from the insects, log the extent of their decline, and then reverse and/or nullify the decline by vitamin K replacement. GREEN (1963) states that vitamin A in large doses seems to be a vitamin K antagonist. He cites the work of MOORE and WANG (1945) who produced hemorrhages in rats overdosed with vitamin A, and LIGHT *et al.* (1944) who demonstrated a severe hypothrombinemia after giving rats 18,000 μg. of vitamin A per day for ten days, which could be prevented with as little as 25 μg. of vitamin K_1. Although the mechanism of the antagonism is obscure, it has been suggested by QUICK and STEFANINI (1948) that large amounts of vitamin A interfere with the synthetic functions of rat intestinal flora.

a) Reversing vitamin A-induced aposymbiotism with vitamin K

The parallelism existing between insects and warm-blooded animals becomes more apparent as more is learned about the chemistry and physiology of insects. The role of vitamin A in suppressing yields of vitamin K

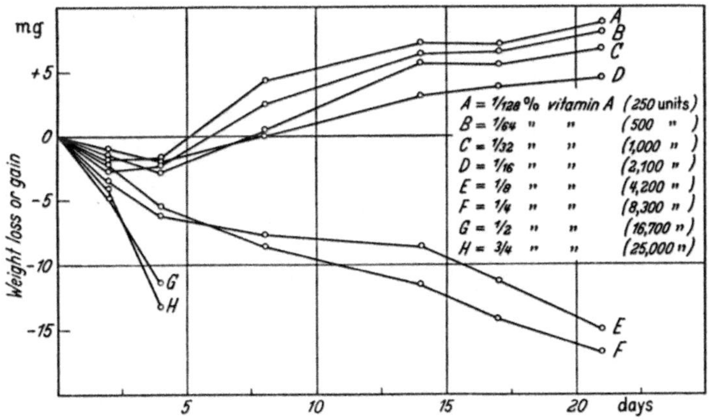

Fig. 8a. Effects of various concentrations (units and percentage) of vitamin A (Aquasol A, 25,000 USP units per ml.) in fabric on half-developed furniture carpet beetle larvae during 21 days exposure. The breaking point between balance and imbalance of vitamin A lies somewhere between 1/16 and 1/8 percent. Rapid weight loss and mortality at 1/2 and 3/4 percent vitamin A is in part due to excessive "oiliness" at these concentrations

at the symbiont level in rats is now suspected of being the case with the two species of carpet beetle larvae used for these studies. Excessive doses of vitamin A will not only cause the insects to decline via possible apo-

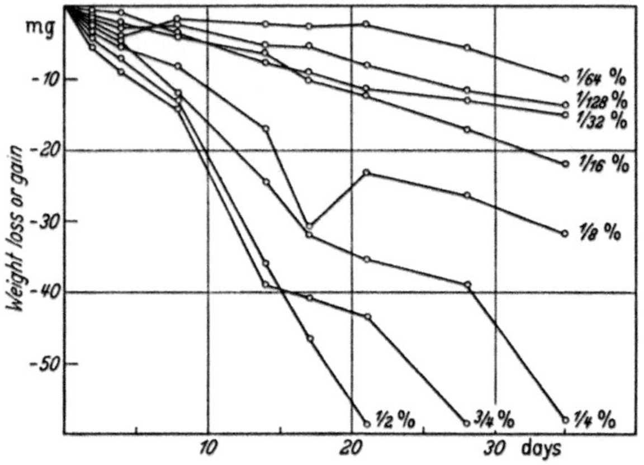

Fig. 8b. Effects of various concentrations (same units as in Fig. 8a) of vitamin A (Aquasol A, 25,000 USP units per ml.) in fabric on half-developed black carpet beetle larvae during 35 days exposure. Black carpet beetle larvae are more adversely sensitive to vitamin A than the furniture carpet beetle. Rapid weight loss and mortality at $^1/_4$, $^1/_2$ and $^3/_4$ percent as also due in part to excessive "oiliness" at these concentrations

symbiotism (see Figs. 8a and b) but trace amounts of vitamin K administered orally will significantly reverse this deficiency (see Figs. 9 and 10).

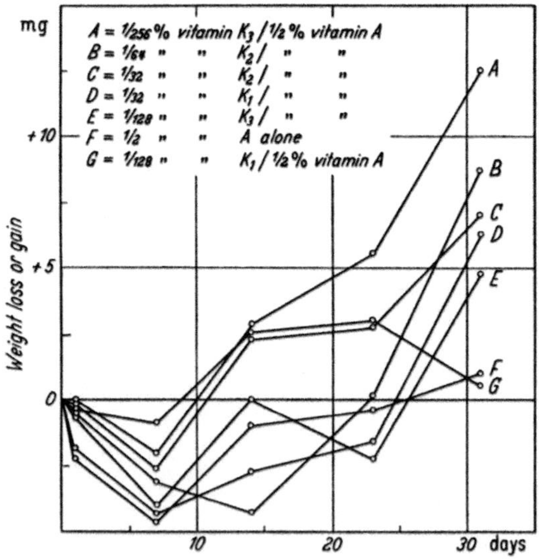

Fig. 9. Reversal effects in 31 days of various percentages of vitamins K_1, K_2 and K_3 for half-developed furniture carpet beetle larvae when K vitamins are added over fabric treated with $^1/_2$ percent vitamin A acetate (1 million units per gram). Furniture carpet beetles have higher immediate acceptance for vitamin A than black carpet beetles. Fatal effects do not occur until prepupal stage is reached. Vitamin A acetate dissolved in acetone

It is interesting to note that normally the black carpet beetle is the most difficult of all fabric-feeding insects to kill via metabolic inhibition where most antimetabolites are concerned. Inasmuch as this is also true when this

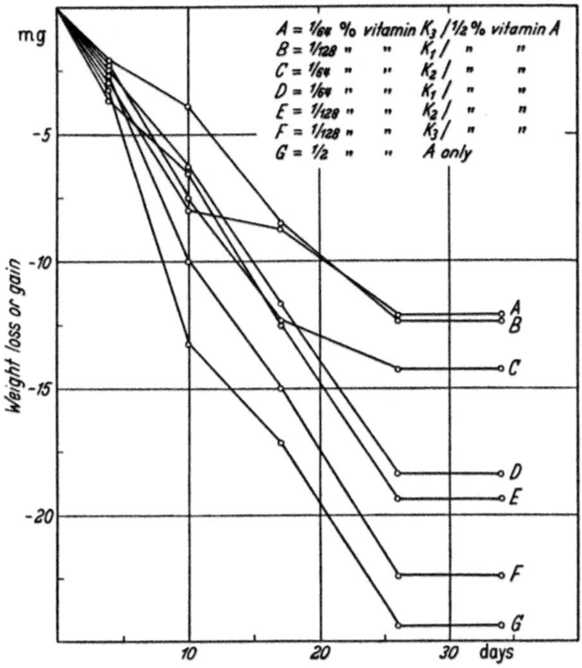

Fig. 10. Partial reversal effects in 34 days of various percentages of vitamins K_1, K_2, and K_3 for half-developed black carpet beetle larvae when K vitamins are added over fabric treated with $^1/_2$ percent vitamin A acetate (1 million units per gram). Vitamin A acetate dissolved in acetone

insect is subjected to direct starvation, it would appear that the species has a more efficient method of making use of its accumulated adipose tissues (Pence 1963). This does not appear to be the case when the black carpet beetle is subjected to excessive intake of vitamin A. Early symptoms of decline then occur quickly and continue if not arrested by vitamin K replacement. Apparently the black carpet beetle has a greater need for vitamin K than the furniture carpet beetle.

b) Vitamin K deficiency transmitted to offspring

Vitamin A affects the furniture carpet beetle more slowly than the black carpet beetle. It also manifests a slightly different mode of action, for "sub-lethal" doses may not take their toll until the last instar when the insect dies rather suddenly and without the same noticeable "dehydration" found in the black carpet beetle. Current studies are underway whereby vitamin A is being further reduced in the furniture carpet beetle diet which enables the larvae barely to reach a vitamin K-deficient state of maturity and reproduce. To date it has been observed that most of the first instar larvae die prior to the first molt. While more study is required, it is interesting to speculate on the possibility of transmission of vitamin K deficiency to offspring. BROOKS (1963) discusses the elaborate systems that exist in some insects for the transmission of symbiotic microorganisms to their offspring, the modes of transmission as related to the definitive location of the symbionts in the adult insect, and to the trophic structure of the ovary. The history of modern medicine reveals that, prior to present knowledge of the importance of vitamin K to human beings, it was not uncommon for infants to die immediately following birth due to a condition which was to be diagnosed later as hypoprothrombinemia caused both by vitamin K depletion in the mother during gestation and a fatal lack of vitamin K in the new-born infant. This condition is now prevented by administering vitamin K to the mother before parturition or by giving the infant a small dose of the vitamin.

In testing for the biological effects of vitamin K inhibition, certain important factors relative to the basic diet of test insects used for these studies must be taken into account. The furniture carpet beetle will readily accept a diet of wool upon which it has been reared. The black carpet beetle, however, is reared on ground dog food, after which it is transferred to wool test cloth for mothproofing evaluation, in compliance with CSMA standards. This transition, however, is always too abrupt, for even when test wool is fortified it will not be accepted immediately. In due time noticeable feeding occurs which is further evidenced by substantial excrement weight yields, yet still the insect's body weight remains static and, in fact, may experience some loss before gain begins. Obviously this factor should be taken into consideration when attempting dietetically to inhibit an unaccustomed diet of wool, as well as during any experiment to reverse the inhibition with the very nutrient in which the test insect becomes deficient. Economic loss because of insect damage to wool is of primary concern, and it becomes necessary therefore that this material be the standard substrate upon which all proofing tests are made.

Summary

Control of fabric-feeding insects with metabolic inhibitors necessitates certain changes in existing CSMA standards for evaluating proofing effectiveness. Inasmuch as non-pesticidal agents are used, body weight losses are substituted for mortality. Twenty-five rather than ten insects are employed and ultimate proofing effectiveness is measured by lack of fiber weight loss.

Use of antimetabolites and excesses of essential nutrients affords a safe approach to pest control. In the main, metabolic inhibitors resemble chemical structures of metabolites and are considered relatively non-toxic. Based on mg./kg. of body weight, only minute amounts of these inhibitors are required to disrupt the metabolism of insects. Proportionately, certain nutrients in amounts beneficial to man can induce symptoms of hypervitaminosis in insects. Thus the opposite ends of the nutrient requirement spectrum become areas of investigation in pest control that contribute much toward the elimination of toxic residues normally encountered through use of conventional pesticides.

The inhibition and/or excessive use of vitamin K will control carpet beetles. Dietetic studies have shown these insects to have a need for vitamin K which is fulfilled biosynthetically as well as orally. Vitamin K is not normally associated with processed wool and must therefore be made available through symbiotic microorganisms; yet, in spite of this supply, trace amounts of orally administered vitamin K will promote growth and development. Conversely, amounts in excess of trace are deleterious and induce mortality.

Vitamin K is inhibited by both dicumarol and salicylic acid. Both inhibitors can be reversed if vitamin K is replaced before irreparable damage occurs. The biochemical role of vitamin K in fabric-feeding insects is not fully understood. With 420 species of insects known to have coagulable hemolymph, it is conceivable that vitamin K deficiency in these insects may interfere with clotting mechanisms and lead to prolonged coagulation time or a syndrome may be involved between anticoagulation and interference with oxidative phosphorylation.

Aposymbiotism is believed to be induced in carpet beetle larvae by excessive intake of vitamin A, resulting in a fatal vitamin K deficiency. However, early symptoms of this deficiency can be reversed if minimal amounts of vitamin K are added to the diet. It is thus shown that lethal hypervitaminosis can be induced directly in carpet beetle larvae by overbalancing an essential nutrient while indirect lethality can occur when nutrient excesses eliminate metabolite-producing microorganisms within the gut. Artificially induced aposymbiotism also appears to contribute toward transmission of vitamin K deficiency to offspring. Sublethal doses of vitamin A in the carpet beetle diet enable the insects to mature and reproduce; however, preliminary studies indicate that first instar progeny are incapable of completing the first molt. Additional work in this interesting area of investigation is currently under study, as well as continued examination of the remaining fat-soluble vitamins. The relationship between vitamin K and vitamin E is being further explored, for preliminary results indicate the inhibition of vitamin E interferes with the carpet beetle's reproductive system.

Résumé *

La lutte contre les insectes que se nourrissent d'étoffes, à l'aide d'inhibiteurs de métabolisme, exige certaines modifications des standards CSMA existant pour mesurer l'efficacité de l'épreuve. Lorsque l'on utilise des produits non pesticides, les pertes de poids corporels sont substituées à la mortalité. On utilise vingt cinq insectes plutôt que dix et l'épreuve finale d'efficacité est basée sur l'absence de perte de poids de fibres.

L'utilisation d'antimétabolites et d'excès de substances nutritives essentielles constituent une voie sûre dans la lutte contre les insectes. En général,

* Traduit par S. Dormal-van den Bruel.

les inhibiteurs de métabolisme ont une structure chimique semblable à celle des métabolites et sont considérés comme relativement non toxiques. Pour rompre le métabolisme des insectes, il ne faut que des quantités infimes de ces inhibiteurs, de l'ordre du mg/kg de poids corporel. Toutes proportions gardées, certaines substances nutritives en doses utiles pour l'homme peuvent donner lieu à des symptômes d'hypervitaminose chez les insectes. Ainsi, les extrêmes opposés de la gamme des besoins en substances nutritives deviennent des domaines de recherche dans la lutte contre les insectes qui contribuent beaucoup à eliminer les résidues toxiques apportés normalement par l'usage des pesticides conventionnels.

L'inhibition et/ou l'application excessive de vitamine K détruira les anthrènes. Des études diététiques ont montré que ces insectes ont un besoin en vitamine K qui est couvert par biosynthèse autant que par ingestion. La vitamine K n'est normalement pas associée à la laine apprêtée doit donc être fournie par des micro-organismes symbiotiques; cependant, malgré cet apport, des traces de vitamine K administrées par voie orale favoriseront la croissance et le développement. Par contre, des doses excédant les quantités infimes sont nocives et induisent la mortalité.

La vitamine K est inhibée par le dicoumarol et l'acide salicylique. Ces deux inhibiteurs peuvent être neutralisés si la vitamine K est réintroduite avant l'apparition de dégâts irréparables. Le rôle biochimique de la vitamine K chez les insectes qui se nourrissent de'étoffes n'est pas complètement élucidé. Etant donné qu'il y a 420 espèces d'insectes connus comme ayant une hémolymphe coagulable, il est concevable qu'une déficience en vitamine K chez ces insectes quisse interférer dans les mécanismes de coagulation et prolonger le temps de coagulation ou qu'un syndrôme soit déclanché entre l'anticoagulation et l'interférence avec la phosphorylation oxydative.

On croit qu'une absorption excessive de vitamine A induit, chez les larves d'anthrènes, un aposymbiotisme qui entraîne une déficience léthale en vitamine K. Cependant, les premiers symptômes de cette déficience peuvent être neutralisés si des quantités minimes de vitamine K sont ajoutées au régime. Il est ainsi démontré qu'une hypervitaminose léthale peut être induite directement chez les larves d'anthrènes par excès de substances nutritives essentielles, tandis qu'une mortalité indirecte peut survenir lorsque des excès de substances nutritives éliminent les micro-organismes produisant les métabolites dans l'intestin. L'aposymbiotisme induit artificiellement semble aussi contribuer à transmettre les déficiences en vitamine K à la descendance. Des doses sub-léthales de vitamine A dans le régime de l'anthrène permettent aux insectes d'atteindre le stade adulte et de se reproduire; cependant, des études préliminaires indiquent que les larves du premier âge sont incapables d'accomplir leur première mue. Des travaux complémentaires dans cet intéressant domaine de recherches sont en cours, de même qu'en ce qui concerne les autres vitamines solubles dans les graisses. La relation entre la vitamine K et la vitamine E est examinée de plus près, car les résultats préliminaires indiquent que l'inhibition de la vitamine E interfère dans le processus de reproduction chez l'anthrène.

Zusammenfassung *

Bei der Bekämpfung der textilzerstörenden Insekten mit Stoffwechsel-Inhibitoren ergibt sich die Notwendigkeit, die bestehenden CSMA-Standards für die Auswertung von Schutztesten abzuändern. Da nicht pestizid wirksame Agentien benutzt werden, tritt an die Stelle der Mortalität der Verlust an Körpergewicht. An Stelle von 10 Insekten sind 25 Versuchstiere zu benutzen; die letzte Beurteilung der Schutzwirksamkeit eines Präparates erfolgt durch Fehlen des Faser-Gewichtsverlustes.

Die Benutzung von Antimetaboliten und die Überfütterung mit lebensnotwendigen Nährstoffen bedeutet einen gefahrlosen Weg zur Schädlingsbekämpfung. Inhibitoren des Stoffwechsels ähneln den Metaboliten in der chemischen Struktur und werden als verhältnismäßig ungiftig betrachtet. Berechnet auf mg/kg Körpergewicht sind nur sehr geringe Mengen dieser Inhibitoren notwendig, um den Stoffwechsel der Insekten irreversibel zu stören. Bestimmte Nährstoffe verursachen in Aufwandmengen, die für den Menschen günstig sind, bei Insekten alle Symptome der Hypervitaminose. Auf diese Weise werden die verschiedenen Aspekte des Nährstoffbedürfnisses zu Untersuchungsgebieten der Schädlingsbekämpfung; sie tragen dazu bei, daß toxische Rückstände, wie sie bei der Anwendung konventioneller Pestizide auftreten, vermieden werden.

Durch Hemmung von und/oder Überfütterung mit Vitamin K kann eine wirksame Kontrolle der Teppich-Käfer erreicht werden. Ernährungsphysiologische Untersuchungen haben gezeigt, daß die Insekten einen Vitamin K-Bedarf haben, der sowohl durch orale Aufnahme als auch biosynthetisch gedeckt wird. Vitamin K ist normalerweise in verarbeiteter Wolle nicht vorhanden und muß daher durch symbiontische Mikroorganismen zugänglich gemacht werden. Trotz dieser Zufuhr fördern kleine Mengen von oral gegebenem Vitamin K das Wachstum und die Entwicklung. Hohe Dosen sind hingegen schädlich und induzieren Mortalität.

Vitamin K wird sowohl durch Dicumarol als auch Salicylsäure gehemmt. Die Wirkung beider Hemmstoffe kann jedoch aufgehoben werden, wenn Vitamin K gegeben wird, bevor irreversible Schäden eingetreten sind. Die biochemische Bedeutung des Vitamin K in textilzerstörenden Insekten ist noch nicht völlig verständlich. Bei 420 Insekten-Arten, von denen bekannt ist, daß sie koagulierbare Hämolymphe besitzen, kann angenommen werden, daß Vitamin K-Mangel beim Koagulationsmechanismus eine Rolle spielt. Es führt zu einer verlängerten Koagulationszeit oder zu einem Syndrom, das einbezogen ist zwischen Antikoagulantien und Interferenz mit der oxydativen Phosphorylierung.

Es wird angenommen, daß exzessive Aufnahme von Vitamin K bei den Larven des Teppich-Käfers Aposymbiotismus induziert, der zu einem fatalen Vitamin K-Mangel führt. Frühe Symptome dieses Mangels können jedoch durch kleinste Mengen von Vitamin K in der Nahrung aufgehoben werden. Auf diese Weise kann gezeigt werden, daß letale Hypervitaminose in Larven direkt induziert werden kann durch Überfütterung mit einem lebensnotwendigen Nährstoff, während direkte Letalität vorkommen kann, wenn

* Übersetzt von H. F. Linskens.

durch Überfütterung die metabolitenproduzierenden symbiontischen Mikroorganismen im Verdauungstraktus zum Verschwinden gebracht wurden. Künstlich induzierter Aposymbiotismus kann zur Übertragung des Vitamin K-Mangels auf die Nachkommen führen. Subletale Dosen von Vitamin A im normalen Substrat ermöglichen es dem Insekt, die volle Reife zu erreichen und sich fortzupflanzen; vorläufige Untersuchungen zeigten jedoch, daß die 1. Larvalgeneration nicht in der Lage ist, die 1. Häutung durchzuführen. Weitere Untersuchungen auf diesem interessanten Gebiet sind, ebenso wie die Prüfung der anderen fettlöslichen Vitamine, im Gange. Die Beziehung zwischen Vitamin K und Vitamin E wird untersucht. Die vorläufigen Ergebnisse zeigen jedoch bereits, daß die Hemmung des Vitamin E zu einer Störung des Fortpflanzungssystems der Teppich-Käfer führt.

References

Beard, R. L.: Experimental observations on coagulation of insect hemolymph. Physiol. Zool. **23**, 47 (1950).

Best, H. C., and N. B. Taylor: The physiological basis of medical practice. 7 ed. Baltimore: Williams & Wilkins 1961.

Block, R. J.: Chemical classifications of keratins. Annals N. Y. Acad. Sci. **53**, 608 (1951).

Brooks, M. A.: Symbiosis and aposymbiosis in arthropods. Symbiotic associations. Cambridge: Cambridge Univ. Press 1963.

Gilmour, D.: Biochemistry of insects. New York-London. Academic Press 1961.

Green, J.: Antagonists and inhibitors of the fat-soluble vitamins. Metabolic inhibitors. Vol. 1. New York-London: Academic Press 1963.

Gregoire, R. C.: Reaction of insect hemolymph to vertebrate coagulation inhibitors. Biol. Bull. **104**, 372 (1953).

— Blood coagulation in arthropods. Arch. Biol. **66**, 104 (1955).

Gustaffson, B. E., F. S. Daft, E. G. McDaniel, J. C. Smith, and R. F. Fitzgerald: Effects of vitamin K-active compounds and intestinal microorganisms in vitamin K-deficient germ-free rats. J. Nutrition **78**, 461 (1962).

Harper, H. A.: Review of physiological chemistry. 7th ed. Los Altos, Calif.: Lange Med. Publications 1959.

Light, R. F., R. P. Alscher, and C. N. Frey: Vitamin A toxicity and hypoprothrombinemia. Science **100**, 225 (1944).

Moore, T., and Y. L. Wang: Hypervitaminosis A. Biochem. J. **39**, 222 (1945).

Pence, R. J.: A new approach to fabric pest control. Soap & Chem. Specialties **35**, 65 (1959).

— The antimetabolite imidazole as a pesticide. J. Econ. Entomol. **65**, 1 (1963).

— Use of antimetabolites for pest control. 2 ed. Encyclopedia of Chemistry. New York: Reinhold 1965.

Quick, A. J., and M. Stefanini: Experimentally induced changes in the prothrombin level of the blood. J. Biol. Chem. **175**, 945 (1948).

Wooley, D. W.: A study of antimetabolites. New York-London: Wiley 1952.

Wright, S.: Applied physiology. 9 ed. London-New York-Toronto: Oxford Univ. Press 1956.

Antibiotics in foods — naturally occurring, developed, and added

By

ELMER H. MARTH*

Contents

* Research and Development Division, National Dairy Products Corporation, Glenview, Illinois.

I. Introduction

The presence of antibiotics in certain foods, especially milk, poultry, and fish, has been the subject of much research and discussion. Antibiotics, in these instances, were added to the foodstuff purposefully to preserve it or accidentally through the treatment of diseased animals. Inhibitory substances may also appear in foods in two other ways. They may occur naturally in foods as has been noted with garlic, onion, horseradish, banana, and other foods, or they may be produced in a food during its manufacture. Examples of the latter are certain fermented milks and cheeses.

The purpose of the present paper is to summarize information on the presence, in foods, of antibiotics from these three principal sources. The discussion will first consider naturally occurring antibiotic substances in foods, then inhibitory substances developed in foods during their production, and finally antibiotics added to foods. Methods for detection of these residues of inhibitory substances and regulations of various countries affecting the addition of antibiotics to foods will not be considered in detail. Some of the residue and residue-analytical problems associated with certain antibiotics in certain foodstuffs and feeds have recently been reviewed elsewhere (Marth 1961 b, Allen and Beckman 1964, Myers 1964, Katz and Helrich 1964). Comprehensive reviews of methods for detection and determination and of international regulations of these important substances are needed.

II. Naturally occurring antibiotic substances

Antibiotic substances have been recovered from many naturally occurring substrates. Included are plants of many types, seeds, bulbs, honey, fresh raw milk, and other materials. Screening tests have been conducted on many "exotic" plants in the hope of finding antimicrobial agents which might be of use medically or in the preservation of foods. The discussion to follow will be concerned principally with those plants and other materials which may find their way into foods or feeds.

a) Onion

The fresh juice of onions was reported by Pasteur, in 1857, as able to inhibit the growth of beer yeasts (see Bustinza 1959). In 1937 Lovell observed that vapors from crushed onions inhibited *Bacillus subtilis* and *cereus* var. *mycoides, Serratia marcescens,* and *Mycobacterium smegmatis.* Emission of bactericidal vapors was enhanced by increasing the temperature of the crushed onion from 10° to 37° C. Onion juice was tested for its inhibitory properties by Fuller and Higgins (1940) who noted that *Aerobacter aerogenes* was not inhibited by the juice but the following organisms were retarded by various concentrations: *Escherichia coli, Salmonella schottmuellerei, Pseudomonas pyocyanea, Staphylococcus aureus, Proteus vulgaris,* and bacilli including *B. subtilis, cereus,* and *megaterium,* and others. Seitz filtration and steaming reduced antibiotic potency of the juice but a pH between 5.8 and 8.0 had no effect.

Tests conducted more recently by Bustinza (1959) verified some of the earlier observations. He noted that fresh onion juice inhibited the following yeasts: *Saccharomyces cerevisiae, cerevisiae* var. *ellipsoides,* and *oviformis, Candida albicans, Torula major, Kloechera apiculata,* and *Hansenula anomala.* Growth of the fungus *Tricophyton mentagrophytes* was also retarded.

Virtanen and Matikkala (1959 a) found that growth of *S. aureus* was stopped completely when 50 to 100 mg. of crushed onion was added per ml. of culture. Fresh and heated onions were tested by Wilson and Brown (1953) who found these preparations inhibited *S. aureus, B. subtilis* and *thermoacidurans, Acetobacter aceti, Leuconostoc mesenteroides,* and *Lactobacillus plantarum* and *lycopersici.*

Virtanen and Matikkala (1959 b) found that fresh homogenized onions have a strong antimicrobial effect and that growth of *S. aureus* was stopped completely by addition of 100 mg. of crushed onion/ml. of nutrient solution. Addition of 15 mg. retarded growth but failed to inhibit the bacterium. They observed further that heating to boiling or freezing with Dry Ice reduced the inhibitory activity so that the presence of 1.0 g. of onion/ml. had no effect on growth of *S. aureus.* These authors recovered

S-methylcysteine-sulfoxide and *S*-*n*-propylcysteine-sulfoxide from onions. Both compounds are, through enzymatic action, converted to the corresponding thiosulfinates which exhibit strong antimicrobial action against many microorganisms. The effect of these compounds appears similar, though somewhat weaker, to that of allylthiosulfinate formed from *S*-allylcysteine-sulfoxide present in garlic. Since a reaction catalyzed by enzymes is necessary for development of the antibiotic substances in onions, it is readily understandable why exposing them to treatments as rigorous as boiling would eliminate most of the inhibitory activity.

Other compounds identified as present in onions include: methanol, ethanol, acetaldehyde, propionaldehyde, and 2-methyl-2-pentenal (SPARE and VIRTANEN 1961); cycloalliin (VIRTANEN and MATIKKALA 1956, 1958, and 1959 a; MATIKKALA and VIRTANEN 1957); and a variety of peptides inculding γ-glutamyl-L-phenylalanine and γ-glutamyl-*S*-β-carboxy-β-methyl-ethyl-cysteinyl-glycine (VIRTANEN and MATIKKALA 1960 and 1961). No special antibiotic action has been associated with any of these substances.

b) Garlic

Many therapeutic virtues have been ascribed to garlic and its effect on growth of microorganisms has been studied for some time. WALTON *et al.* (1936) found that garlic vapors were inhibitory to *Mycobacterium butyricum* and *smegmatis* and to a lesser extent to *E. coli, S. marcescens,* and *B. subtilis.* The active substance was more volatile at 37.5° C. than at 10° C. and was destroyed by boiling and autoclaving. WILSON and BROWN (1953) examined heated and unheated garlic and found these preparations able to inhibit *S. aureus, B. subtilis* and *thermoacidurans, E. coli, A. aceti, Clostridium sporogenes, Saccharomyces ellipsoideus, L. mesenteroides,* and *L. plantarum,* and *lycopersici.*

The principle substance responsible for the antimicrobial action of garlic is called "allicin" and has been reported as highly inhibitory to gram-positive and -negative bacteria, some fungi, and *Mycobacterium tuberculosis* (RAO and VERMA 1952). The same authors indicate that initially allicin was believed to have a diallyl thiosulfinate structure but their studies showed it to be diallyl disulfide oxide. Derivatives were prepared and diphenyl disulfide oxide proved more inhibitory to *S. aureus, E. coli, Aerobacter aerogenes,* and four *Salmonella* spp. than did *p*-acetamino-*p*-nitrodiphenylthiolsulfonate, sodium *p*-acetamino-phenylthiolsulfonate, (*p*-carboxyphenyl) ethylxanthate, and *o*-tolylethylxanthate. Several γ-glutamyl-peptides have been recovered from garlic by VIRTANEN and MATTILA (1961 a and b), and by VIRTANEN *et al.* (1962). One, γ-L-glutamyl-*S*-allyl-L-cysteine, is believed to be the precusor of *S*-allylcysteine sulfoxide, the compound which may give rise to the actual inhibitory agent. DATTA *et al.* (1948) dialyzed the whole clove of garlic with ether for six hours and after further purification obtained two active fractions. One was equally inhibitory to *S. aureus* and *E. coli* while the other was most active against the staphylococcus. Their fractions were insoluble in water whereas allicin is soluble; thus garlic may contain several antibacterial agents.

The juice of garlic cloves and aqueous as well as certain organic solvent extracts of commercial garlic powder inhibited bacterial and fungal plant pathogens according to ARK and THOMPSON (1959). They observed further that downy mildew of cucumber and of radish, cucumber scab, bean rust, bean anthracnose, early blight of tomato, brown rot of stone fruits, angular leaf spot of cucumber, and bacterial blight of beans were effectively controlled by sprays containing one-to-20 percent aqueous extracts of garlic powder preparations. A dust consisting of garlic powder and a carrier was found able to control downy mildew of cucumber and bean rust. The pungent odor of garlic preparations was neutralized with α-neutroleum and the resulting materials were pleasantly aromatic. Garlic powders were found to retain full activity for at least three years when kept dry.

CAVALLITO (1951) patented a process for the recovery of allyl disulfide oxide from garlic cloves. His preparation was claimed to be 400 times more potent than ground garlic cloves and when tested against *S. aureus*, 1.0 mg. had activity equivalent to 15 Oxford units of penicillin. The process employed an initial extraction of cloves with ethyl alcohol, followed by removal of the alcohol below 50° C., co-distillation at reduced pressure with water below 50° C., extraction of the oxide from the distillate with ethyl ether, and finally removing the ether by distillation.

The use of hydrogen peroxide to stabilize the antibiotic extracted from garlic was reported by LEJNIEKS (1956). Peroxide was added to an alcohol extract of cloves, allowed to react, and any excess neutralized with manganese dioxide. Maximum activity remained in the preparation if the solution was at pH 5.0 to 6.0 during treatment with peroxide.

Preparation of garlic powders with maximum antibiotic activity was investigated by PRUTHI *et al.* (1959 a, b, c, d, and e). Their tests indicated a) temperatures during drying of garlic should not exceed 60° C., b) the optimum tray load during drying is 1.14 pounds of freshly prepared garlic per square foot of tray surface, c) sulfuring or sulfiting of garlic before dehydration improved the color of garlic powder but significantly reduced its flavor and completely destroyed its antibacterial activity, d) addition of ascorbic acid had no beneficial effect on retention of antibacterial activity or flavor, e) freeze-drying and vacuum-shelf drying produced a product with improved color compared to that obtained by hot-air or sun drying although there was no significant difference between methods in retention of flavor or antibiotic activity, and f) loss of antibacterial activity in powder was reduced by storing under nitrogen and at 0° to 2° C. DUBROVA (1956) has verified that retention of antimicrobial properties in garlic is enhanced by low temperature storage.

c) Horseradish

HALBEISEN (1957) extracted horseradish with organic solvents and recovered an essential oil which contained an antibiotic substance. The oil was diluted with Tween 80 in preparation of a stock solution which contained 500 mg. per ml. Both gram-positive and -negative bacteria, including *S. aureus, Streptococcus faecalis, E. coli, P. vulgaris, Salmonella typhosa, Klebsiella pneumoniae, B. subtilis* and *anthracis, Pseudomonas aeruginosa,*

Corynebacterium diphtheriae, Listeria monocytogenes, and others, were inhibited by the presence of 1.5 to 100 mg. of extracted essential oil per ml. of culture medium. Bactericidal doses were ten times greater for gram-positive and 25 times greater for gram-negative organisms than were requirements for bacteriostasis. Resistance of bacteria to the antibiotic increased only slightly even after repeated transfer. The antibiotic also proved inhibitory to a variety of pathogenic fungi included in the genera *Trichophyton, Trichosporon, Microsporon, Epidermophyton, Achorion, Sporotrichon, Histoplasma, Nocardia, Blastomyces,* and *Candida.* Activity of the antibiotic was not affected by changing the electrolytes of the solution nor by adding reducing substances.

The LD_{50} of the antibiotic, determined in white mice, was found to be 160 mg. per kg. When the preparation was ingested, it caused a mild burning sensation on the tongue. Administering it to damaged skin was accompanied by a warming sensation but no other effects. Although ingestion of fresh horseradish roots had no demonstrable effect on the intestinal flora, consumption of the oil contained in a capsule able to dissolve in the small intestine was accompanied by a disturbance of short duration, not exceeding 24 hours. WILSON and BROWN (1953) tested heated and unheated horseradish for its antibacterial activity and found the following organisms were inhibited by the preparations: *S. aureus, B. subtilis* and *thermoacidurans, A. aceti, C. sporogenes,* and *L. mesenteroides.*

d) Banana

The pulp and skins of green, naturally ripened, and ethylene-ripened bananas, banana leaves, and petioles were examined for antibiotic substances by SCOTT *et al.* (1949). They extracted fresh plant material with water while dried material, after grinding, was extracted with methanol or petroleum ether. All extracts were concentrated so each ml. of solution was equivalent to one g. of plant material. Solvents were replaced by water and solutions were adjusted to pH 6.0 to 7.0 before testing.

Cylinder cup-plate and disc assay methods (see KATZ and HELRICH 1964) were used to evaluate the solutions. Extracts of banana leaves were found to be only slightly antibacterial, whereas extracts of stems inhibited some gram-positive and acid-fast bacteria. The fungus *Fusarium oxysporum* was inhibited by extracts of the pulp and skin of green bananas while similar extracts from ripened bananas effectively inhibited *F. oxysporum, E. coli, S. aureus, S. marcescens, Mycobacterium phlei, B. subtilis, Sarcina lutea, Rhodococcus roseus,* and *Xanthomonas translucens.* Distillates recovered from ripened banana skins were also found inhibitory to a similar group of microorganisms. The authors speculated that since green bananas yielded an antifungal agent and ripe bananas an antibacterial substance, at least two separate antibiotic agents are associated with bananas and the antibacterial material is formed during ripening.

e) Sweet potato

BRUCKNER *et al.* (1949) examined various portions (including tubers) of the sweet potato plant for the presence of antibiotic substances. They used *Fusarium*-wilt-resistant and -susceptible sweet potato varieties. Hot and cold aqueous extracts failed to inhibit test microorganisms and, in most instances, actually stimulated their growth. Methanolic extracts of stems, when concentrated to dryness and diluted with water so that one ml. contained the extract of one g. of material, inhibited *F. oxysporum, E. coli, S. aureus, S. lutea, S. marcescens, M. phlei, B. subtilis* and *cereus, X. translucens,* and *R. roseus.* Roots and tubers also contained the inhibitory agent but leaves and stems proved to be the best source. The authors were able to separate the antimicrobial agent into components and concluded the sweet potato plant contained at least three discrete biologically active materials.

f) Cabbage

The juice of cabbage was reported (SHERMAN and HODGE 1936) to contain a mild bactericidal agent more active than that of freshly drawn milk but less than that found in fresh normal blood.

Exposure of juice to 60° C. for ten minutes or Berkefeld filtration eliminated the inhibitory activity. Cabbage juice was able to inhibit *E. coli, A. aerogenes,* and *Pseudomonas campestre.* Additional studies on cabbage juice by PEDERSON and FISHER (1944) showed that the inhibitory substance was most active against gram-negative bacteria such as those in the genera *Achromobacter, Flavobacterium,* and *Pseudomonas,* and also *E. coli* as well as the gram-positive bacterium, *S. aureus.* Differences in concentration of the active substance were associated with several cabbage varieties.

LITTLE and GRUBAUGH (1946) observed one variety of cabbage contained a substance inhibitory to *S. typhosa* but not to *S. aureus, Salmonella paratyphi,* or *E. coli.* Other varieties of cabbage examined failed to inhibit any of the test bacteria but one retarded growth of *F. oxysporum.* Heated and unheated cabbage, according to WILSON and BROWN (1953), inhibited *B. subtilis* and *A. aceti.*

Attempts by DICKERMAN and LIBERMAN (1952) to elucidate the chemical nature of the inhibitory substance associated with cabbage revealed that the material was a carbohydrate with a molecular weight of 10,000 or less and that it inhibited *Streptococcus pyogenes, S. aureus, S. typhosa, E. coli, B. cereus* var. *mycoides,* and *B. subtilis* and *coagulans.*

g) Radish

An antibiotic substance was recovered from seeds of radish by IVANOVICS and HORWATH (1947). The material, named "raphanin", was active against *S. aureus* and *E. coli.* WALDER (1952) reported that this agent also inhibited *S. typhosa, P. pyocyanea,* and *B. anthracis* and *subtilis,* and prevented germination of certain seeds.

Raphanin, according to IVANOVICS and HORWATH (1947), when freshly distilled had a slightly yellowish color which darkened on storage at room temperature. The substance had a radish-like odor, was soluble in water, was levorotatory in ethanol, gave no color reaction with ferric chloride or nitroprusside, turned purple with nitroprusside after prior treatment with hydrochloric acid and zinc dust (indicating —SH groups), and formed a heavy precipitate when heated with silver or lead salts. Raphanin was rapidly inactivated at an alkaline pH and by hydrogen sulfide. The substance was found toxic to mice at levels of seven mg. or more and to guinea pigs at levels of 50 mg. or more. Seeds were believed to contain a precursor of the antibiotic which is activated by an appropriate enzyme.

h) Other vegetables

Different investigators have associated antimicrobial activity with a variety of vegetables. WILSON and BROWN (1953) examined several heated and unheated vegetables and concluded that a) *S. aureus* was inhibited by snap beans, beets, white and sweet potatoes, and tomatoes, b) *B. subtilis* was retarded by snap beans, cauliflower, peas, peppers, rhubarb, spinach, and tomatoes, c) growth of *E. coli* was restricted by cauliflower and rhubarb, d) brussels sprouts, cauliflower, and tomatoes inhibited *A. aceti,* f) *B. thermoacidurans* was retarded by cauliflower and rhubarb, g) growth of *S. ellipsoideus* was inhibited by cauliflower, and h) *L. mesenteroides* was inhibited by cauliflower, peppers, and tomatoes. An antifungal agent was isolated from pea pods by CRUICKSHANK and PERRIN (1960). The substance, named "pisatin", was produced by peas as a response to infection by *Sclerotinia fructicola* or other fungal plant pathogens.

MASILUNGAN *et al.* (1959) observed that *M. tuberculosis* was not inhibited by alcohol, water, ether, chloroform, acetone, or acetic acid extracts of lettuce. MASILUNGAN *et al.* (1955) noted that an aqueous but not alcoholic extract of asparagus inhibited *S. aureus* and *E. coli;* they also found that leaves of celery and cucumbers inhibited these two organisms.

Tests with different vegetable juices on *L. plantarum* led PEDERSON and FISHER (1944) to conclude that little if any bactericidal activity was associated with juices of turnip, carrot, cucumber, or Chinese cabbage; cauliflower juice may have displayed some inhibitory action. LABAW and DESROSIER (1953) found that *E. coli* was inhibited by extracts of carrot, green bean, onion, and rhubarb, while *S. aureus* was retarded by the same extracts plus those from tomato, celery, and okra. *B. subtilis* was not inhibited by the okra extract but was sensitive to all the others mentioned previously plus those from chickory, cucumber, sweet corn, and turnip. *S. ellipsoides* was sensitive only to extracts of onion, tomato, and chard.

Antibiotic action of tomatoes has been attributed to the presence of the glycosidal alkaloid tomatine (SEHGAL 1961, CZETSCH-LINDENWALD 1956). This substance has been isolated from both the plant and its roots. Tomatine also contains an aglycone, tomatidine, and a tetrasaccharide moiety consisting of xylose, galactose, and two glucose units.

VALLE (1957) reported that quantitative determinations of chlorogenic and caffeic acids in potato leaves showed that the amount of free chlorogenic acid present is often enough to inhibit mycelial growth of various fungi. Tests with several potato varieties revealed the presence of 0.5 to 4.0 mg. of free chlorogenic acid/g. of fresh leaves. Since much of the chlorogenic acid existed in the bound form, the antifungal effect of the leaves increased when the complex was disorganized. The amount of free caffeic acid present was insufficient to contribute to the total antifungal activity of the leaves.

Seeds of several vegetables were tested by JACOBS and DADD (1959) for inhibitory action against *Corynebacterium fascians*, the causative agent of leafy-gall disease. Carrot, lettuce, corn, parsley, and tobacco seeds exhibited no activity while limited inhibition was associated with radish, wheat, and mustard seeds. Greatest inhibition resulted from broccoli, bean, pea, and tomato seeds, and some antibiotic activity was also demonstrated by clover, flax, and sweet pea seeds.

KESSLER (1955) found that zinc, calcium, iron, and boron ions supplied to potatoes, carrots, and apple trees by means of applied electrical currents of definite strength altered the metabolism of certain nitrogenous compounds in the plant. Intermediates produced from this modified metabolism and isolated from potato tubers, carrots, and apple tree roots exhibited considerable antimicrobial activity against organisms in the genera *Botrytis, Fusarium, Penicillium, Sclerotium, Salmonella, Streptococcus, Bacillus, Agrobacterium, Corynebacterium, Pseudomonas*, and *Xanthomonas*.

i) Fruits

The antibiotic properties of strawberries were investigated by BONDARENKO and ZELEPUKHA (1962) who noted that the berries as well as leaves, flowers, and roots of the plant all contained antimicrobial substances. Although the preparations were slightly toxic some had wound-healing properties.

WILSON and BROWN (1953) examined many fresh and heated fruits for antibacterial activity. They found that *S. aureus* was inhibited by apples, cherries, grape juice, peaches, pineapple, and plums, while growth of *B. subtilis* was retarded by apples, cherries, grape juice, orange juice, pears, pineapple, and plums. Development of *E. coli* was restricted by cherries, grape juice, peaches, pineapple, and plums. *A. aceti* was retarded by apples and grape juice, while apples, cherries, grape juice, pineapple, and plums inhibited *B. thermoacidurans*. Growth of *C. sporogenes* was restricted by pears, pineapple, and plums, while grape juice inhibited *L. mesenteroides* and *L. lycopersici*.

GALLO and VALERI (1954 and 1955) tested the pericarp and seed of avocado for antibacterial activity. Various extracts of the pericarp inhibited *S. aureus* and *S. lutea* but had no effect on *E. coli, S. typhosa*, or *B. subtilis*. Seeds yielded similar results. MASILUNGAN *et al.* (1955) also found that the avocado fruit inhibited *S. aureus* and *E. coli;* bark and roots also retarded *S. aureus* but had no effect on *E. coli*, and neither organism was affected by leaves. The proposed use of the antibiotic from avocados for preservation of foods is discussed elsewhere in this paper.

j) Cereals

VIRTANEN and his co-workers (VIRTANEN *et al.* 1957; VIRTANEN 1958, 1961 a and b) have devoted considerable attention to the presence of antimicrobial substances in cereals, primarily rye. An antifungal factor able to inhibit *Fusarium nivale, Sclerotinia trifoliorum, Penicillium roqueforti, Aspergillus niger,* and *Mucor* spp. was extracted from rye seedlings and identified as 2(3)-benzoxazolinone. HONKANEN and VIRTANEN (1960 a) prepared a series of derivatives from 2(3)-benzoxazolinone and evaluated their antimicrobial activity. Derivatives found most active against *F. nivale* include 3,4-dihydro-2H-1,4-benzoxazine, 2H-1,4-benzoxazine-3(4H)-one, and a modified 4-hydroxy-1,4-benzoxazine compound. Compounds most active against *S. aureus, Pseudomonas fluorescens,* and *E. coli* included modified 4-hydroxy-1,4-benzoxazines: 4-hydroxy-1,4-benzoxazine-2,3-dione and 2,4-dihydroxy-1,4-benzoxazine-3-one. Formation of 2(3)-benzoxazolinone in rye seedlings was studied by HIETALA and VIRTANEN (1960) and VIRTANEN and HIETALA (1960). The glucoside $C_{14}H_{17}O_9N$ in rye seedlings is converted to the aglucone $C_8H_7O_4N$ through enzymatic action. The aglucone is then chemically transformed to 2(3)-benzoxazolinone (HONKANEN and VIRTANEN 1961). HONKANEN and VIRTANEN (1960 b) also recovered linoleic and linolenic acids from rye plants. Their tests suggest that these compounds may contribute to the overall activity against *F. nivale.*

Wheat and corn seedlings were also found to contain an antifungal agent which was identified as 6-methoxy-2(3)-benzoxazolinone by VIRTANEN *et al.* (1957). This compound had antifungal activity similar to that of 2(3)-benzoxazolinone but it was more active against *S. aureus* and *E. coli.*

Seeds of wheat and barley were examined for antibiotic properties by ARK and THOMPSON (1958). Aqueous and methyl alcohol extracts of wheat and barley seeds failed to inhibit gram-negative bacteria but did retard growth of the following gram-positive bacteria: *Corynebacterium michiganense, Micrococcus flavus, B. subtilis,* and *B. cereus* var. *mycoides.* An acid-fast organism, *M. phlei,* was also inhibited, and the antibiotic substance proved active against fungi in the genera *Verticillium, Fusarium, Cladosporium, Aspergillus, Thielaviopsis, Saccharomyces, Monilinia, Glomerella, Uromyces, Pseudoperonospora,* and *Streptomyces.* The antibiotic material was not destroyed by lyophilization, air drying (saturated discs were air-dried for four days), or autoclaving of moistened wheat seed at 15 pounds pressure for ten minutes.

HUMFELD (1947) examined wheat bran for antibiotic properties and recovered a material with characteristics of a fatty acid and able to form a water-soluble potassium salt which had comparatively high activity against *S. aureus, S. faecalis* and *Micrococcus conglomeratus,* but failed to inhibit *E. coli.* HUMFELD (1947) cautioned that inhibitory activity of plant or animal materials used in compounding bacteriological media should not be confused with that of the organism growing on the substrate. Wheat bran has been used as an ingredient in such media. The antibiotic from wheat bran has been suggested for use in prevention of brown spots on stored lemons (FELDMAN 1956).

An antifungal agent was recovered from young barley coleoptile tissue by LUDWIG *et al.* (1960). They found the material able to inhibit *Helminthosporium sativum* and *Monilinia fructicola.* The inhibitory substance appeared in five-day-old tissue but could not be recovered a day later. It is believed to have the formula $C_9H_{17}O_7N$.

k) Spices and essential oils

The value of spices in the preservation and flavoring of foods has long been recognized. Specific factors responsible for their inhibitory action against microorganisms have been recognized more recently.

KATAYAMA and NAGAI (1959) examined nutmeg for volatile components and found fatty acids, phenols, and terpenes. The terpene fraction was broken down further into α-pinene, α-terpineol, linalool, and geraniol. All fractions, at dilutions of 1 : 20 to 1 : 200, inhibited *B. subtilis, Salmonella enteritidis, S. aureus, P. aeruginosa, Proteus morganii,* and *E. coli.* Later KATAYAMA and NAGAI (1960) found the terpenes carvacrol, thymol, isoberneol, vanilline, and salicylaldehyde inhibited all the bacteria listed immediately above at a dilution of 1 : 2,000 or higher. Activity was improved by adding a ketone group to the compounds to form cyclohexanone derivatives, although cyclohexanol derivatives were also inibitory.

CAPEK (1955) reported that the oils recovered from cinnamon, sweet marjoram, thyme, laurel, pimiento, clove, and coriander showed both bactericidal and fungicidal properties. Anise and fennel oils displayed fungicidal activity only, and slight but not significant inhibition was produced by ginger and nutmeg while cumin, pepper, and juniper oils were inactive. Later CAPEK (1956) showed that anethole was responsible for the antifungal action of anise and fennel volatile oils. Russian anise oil was found to contain 87 percent anethole while Czech fennel oil contained 61 percent.

OKAZAKI and OSHIMA (1953) tested the following components of essential oils against pathogenic fungi and various bacteria: borneol, citral, citronellal, citronellol, geraniol, limonene, linalool, menthol and terpineol, and combination of menthol and terpineol. In general they inhibited fungi and were moderately active against the tubercle bacillus but ineffective against *E. coli, S. aureus,* and *Shigella dysenteriae.* Compounds with an aldehyde structure were generally more powerful than corresponding alcohols, especially in α, β-unsaturated substances. Terpenes were equally active in both chain and cyclic forms.

Twenty essential oils were placed 7.0 mm. from cultures of *S. aureus, E. coli, P. aeruginosa,* and *C. albicans* and were allowed to "react" for four hours by KOSCIK (1955). She found that oils of carvom, peppermint, cinnamon, clove, and citrus completely inhibited growth of all test cultures. Thyme, melissa (lemon balm oil), rosemary, valerian, turpentine, sinapine, marjoram, chenopodium, and havandula oils failed to stop growth but retarded it for a few hours. Somewhat less activity was noted with oils of pine, fennel, santal, juniper, and eucalyptus, while oil of anise had no effect.

MARUZZELLA and FREUNDLICH (1959) tested extracts of different spices and found *E. coli* was inhibited by caraway, cardamom, coriander, cumin, fennel, sabadilla, and wormwood; *S. aureus* was retarded by all except sabadilla, cumin, and coriander, and it was also inhibited by anise, celery, dill, and nigella. The pattern for *S. marcescens, M. smegmatis,* and *Erwinia caratovora* was similar to that of the other bacteria. *Streptomyces venezuelae,* however, was most sensitive to spice extracts and its growth was retarded by angelica, anise, asparagus, caraway, cardamon, celery, coriander, cumin, dill, fennel, jambul, mustard, nigella, sabadilla, angostura, and wormwood. The yeast *C. albicans* was most sensitive to angelica, anise, caraway, cardamom, celery, cumin, fennel, sabadilla, angostura, and wormwood.

l) Egg white

Antimicrobial activity has been associated with egg white since 1890 when WURTZ demonstrated it inhibited the anthrax bacillus (see SILVA and BUCKLEY 1962). The inhibitory substance has been identified as conalbumin, a protein which forms a stable pink-colored complex with iron (FEENEY and NAGY 1952).

The influence of conalbumin on growth of different bacteria was investigated by FEENEY and NAGY (1952) who added it to media in sufficient quantity to complex with iron and then provided the amount under test. In general, inhibition was evidenced by an increase in the lag period and a decrease in the growth rate after growth was finally initiated. *Micrococcus lysodeikticus* was inhibited for 15 hours by 1.4 μM./l. of conalbumin and for 82 hours by 4.6 μM./l. *M. conglomeratus* and *B. subtilis* were inhibited by slightly higher concentrations but *P. aeruginosa* and *P. vulgaris* were unaffected at concentrations tested. The authors noted further that an increase in pH of conalbumin or the addition of riboflavin was associated with an increase in its inhibitory activity.

The antifungal activity of egg white was evaluated by SILVA and BUCKLEY (1962) who found that 48 of 50 pathogenic fungi tested were inhibited. Included are organisms in the genera *Actinomyces, Allescheria, Aspergillus, Blastomyces, Candida, Cephalosporium, Cladosporium, Coccidioides, Cryptococcus, Eidamella, Epidermophyton, Emmonsia, Fonsecae, Geotrichum, Histoplasma, Liptosphaeria, Madurella, Microsporum, Nocardia, Paracoccidioides, Penicillium, Phialophora, Piedra, Rhodotorula, Saccharomyces, Sporotrichum, Streptomyces, Torulopsis, Trichophyton,* and *Verticillium.* Concentrations of egg white needed for inhibition ranged from 1.56 to 12.5 mg. of solids/ml. of agar. The inhibitory activity was not affected by storage (a) at room temperature for 30 days, (b) in the refrigerator for six months, or (c) in the freezer for four months. Coagulation, filtration, or foaming of egg white reduced activity. They concluded the antifungal activity was associated with the avidin and/or conalbumin fractions of egg white.

m) Milk

Milk freshly drawn from the normal cow's udder contains several substances able to inhibit different organisms. Coliform bacteria, apparently, are inhibited by two substances, one specific for *E. coli* and another active against other coliforms (BERRIDGE 1955). The latter substance was found to be destroyed by exposure to 55° C. for 30 minutes. Observation of its action on susceptible bacteria revealed first a reduction in motility and then loss of the ability to refract light; the latter change coincided with death of the cells as revealed by plate-count procedures.

Lactenin is the best characterized inhibitor in fresh milk (BERRIDGE 1955). This "substance" has been subdivided electrophoretically into two fractions, L_1 occurring principally in colostrum and L_2 present in milk after heating to 70° C. Pasteurization destroyed 50 percent of L_1 but did not affect L_2. Lactenin has been found able to inhibit *B. subtilis, Streptococcus agalactiae, dysagalactiae,* and *uberis,* and (slightly) *Lactobacillus helveticus. Streptococcus lactis,* the organism commonly used to manufacture many cultured dairy products, was found most sensitive to L_1+L_2, sensitive to L_1, and least sensitive to L_2.

FRANC *et al.* (1958) found milk, whey, and a whey dialysate to be able to inhibit *M. tuberculosis.* Antibacterial activity was associated principally with a hydrophilic fraction but also appeared to a lesser extent in two other fractions.

n) Honey and honey bees

Honey has been used since ancient times to aid in the healing of wounds (WHITE *et al.* 1963). It has been established by different workers that natural, unheated honey shows antibacterial activity. WHITE *et al.* (1963) have reviewed some of the earlier studies. Inhibitory activity of honey, according to DOLD *et al.* (1937) and POTHMANN (1950), has been demonstrated with the following organisms: *E. coli, Salmonella typhi, S. aureus, Salmonella dysenteriae, C. diphtheriae, B. subtilis, B. cereus* var. *mycoides, B. mesentericus,* and *anthracis, Vibrio cholera,* and *M. tuberculosis.* No inhibition of mold or yeast was detected. The inhibitory substance has been designated as "inhibine" (POTHMANN 1950). A method employing an agar-honey mixture and *S. aureus* to measure the concentration of inhibine in honey has been described by DOLD and WITZENHAUSEN (1955). Exposure of honey to temperatures of 50°, 53°, or 56° C. for ten hours reduced but did not eliminate the activity of inhibine according to DUISBERG and WARNECKE (1959), who also observed that storage of honey in the presence of light for five to seven months reduced the activity of inhibine by 45 percent; this light sensitivity had been noted by PRICA (1938).

More recently workers at the Eastern Regional Research Laboratory of the *U.S. Department of Agriculture* have conducted extensive studies on the inhibitory principal in honey. WHITE *et al.* (1963) demonstrated that inhibine is actually hydrogen peroxide produced in honey by the naturally present glucose oxidase acting on glucose. SCHEPARTZ and SUBERS (1964) found the glucose oxidase of honey to be similar to that of honey bees and suggested this may be the source of the enzyme in honey. Other studies by WHITE and SUBERS (1964 a and b) verified the earlier observation that inhibine is heat and light sensitive. Glucose oxidase in honey was found sensitive to photo-oxidation and was destroyed by visible radiation in the 425 to 525 mμ range. Photodestruction of inhibine occurred most effectively at pH 3.0 and was negligible at pH 6.0 to 7.0. A chemical method based on production of color by *o*-dianisidine and peroxidase in the presence of hydrogen peroxide was used by WHITE and SUBERS (1963) to measure the inhibitory activity of honey.

Honey may contain an antibiotic(s) besides hydrogen peroxide. LAVIE (1963) described another inhibitory substance which was photolabile, thermostable at 80° C. for 30 minutes, extractable in hot alcohol, cold acetone, and ether, and volatile at 95° C. Earlier LAVIE (1957 and 1960) had examined honey bees for presence of antibiotic materials. He extracted parts of bees with ethyl alcohol and recovered a thermostable substance which inhibited *Salmonella* spp., *E. coli*, *Proteus* spp., *B. pyocyaneus* and *subtilis*. Maximum activity was found in bees six to 12 days old and the antibiotic was also recovered from bee glue. An antibiotic was also recovered from the cuticular surface, venom, and hypopharyngeal glands of worker bees. Drones, larvae, pupae, and newly emerged bees failed to show inhibitory activity. The antibiotic obtained from bees differed from that recovered by extraction of propolis, beeswax, honey, and royal jelly.

BLUM *et al.* (1959) studied the antibiotic characteristics of the royal jelly of the honey bee and found the inhibitory activity to be associated with 10-hydroxy-Δ^2-decenoic acid, the major component of the lipid fraction. This compound retarded growth of *S. aureus*, *E. coli*, and *Neurospora sitophila*.

o) Antibiotics from natural sources suggested for preservation of foods

JENSEN and co-workers have screened a host of naturally occurring plants for antibiotic substances which might be beneficial in the preservation of foods. A pickling solution fortified with antimicrobial substances recovered from iris bulbs, dry sage, Canadian thistle, *Rhus crenata* (an ornamental shrub), or *Sanguinaria canadensis* (blood-root) was suggested by JENSEN and HESS (1951) for use in the quick-curing of meats. JENSEN (1951 f) described a procedure for recovering the antibiotic from the pits of avocados and found this substance would inhibit growth of *S. aureus* in hams treated with a pickling solution containing a 1 : 5,000 concentration of the material. Canned luncheon meats were also preserved by this substance even when heated to an internal temperature not exceeding 155° F. In further work JENSEN (1951 a) found that the antibiotic recovered from avocado pits could be used to prevent the growth of food poisoning bacteria (e.g., *S. aureus*) in cream-type fillings for pastries.

An antibiotic effective against *S. aureus*, *Bacillus niger*, *C. sporogenes*, *Achromobacter perolens*, *Salmonella typhimurium*, *P. fluorescens*, *Penicillium notatum* and *camemberti*, and *Aspergillus glaucus* was recovered by JENSEN (1951 e) from *Krameria triandra* (Peruvian rhatany) and *Krameria orgenta* (Brazilian rhatany). This substance, too, was effective in the preservation of quick-cured meats and cream-type pastry fillings.

JENSEN and MILLER (1951 c) obtained an antibiotic from grapevines (including the roots) which inhibited *B. subtilis*, *C. sporogenes*, *Brucella suis*, *S. aureus*, and *A. perolens*. Again, this substance was suitable for use in meat pickling solutions. Antibiotics able to inhibit organisms listed above and suggested for use only in meat pickles and cream fillings were also recovered from the broadleaf gum plant (JENSEN 1951 b), pareira *(Chondrodendron tomentosum)* (JENSEN 1951 d), bittersweet *(Celastrus scandens)* (JENSEN and MILLER 1951 a), Osage orange tree *(Maclura pomifera)* (JEN-

SEN 1951 c), common agrimony *(Agrimonia gryposepala)* (JENSEN and MILLER 1951 b), purple prairie clover *(Petalostemum purpurem)* (JENSEN and MILLER 1951 e), spring avens *(Geum vernum)* (JENSEN and MILLER 1951 g), white cedar *(Thuja occidentalis)* (JENSEN and MILLER 1951 h), St. John's wort *(Hypericum perforatum)* (JENSEN and MILLER 1951 f), Iceland moss *(Cetraria islandica)* (JENSEN and MILLER 1951 d), butternut tree *(Juglans cinerea)* (JENSEN and SHERMAN 1951 a), and galingale root *(Alpinia officinarum)* (JENSEN and SHERMAN 1951 b).

p) Antibiotics in other natural substances

An antifungal compound, recovered from red clover *(Trifolium pratense)* by BREDENBERG and HIETALA (1961 a and b), has been designated as trifolirhizin, has the formula $C_{21}H_{24}O_{10}$ with a melting point of 139° to 140° C., and was characterized as a β-glucoside.

DECLERCK (1961) reported that the bitter resins of hops contained antibiotics which may protect beer from certain types of spoilage. These antibiotics, it was suggested, might interfere with the detection of other inhibitory substances fraudulently added to beer.

Anacardic acid (*o*-pentadecadienyl-salicylic acid) occurs in the cashew nut tree (WALDER 1952). The sodium salt of this acid was found to be highly active against all gram-positive bacteria. For example, *S. pyogenes* was destroyed in 15 minutes by a 1 : 200,000 concentration of the compound.

The acacia is cultivated in some areas such as the Philippines principally as a shade tree. Its pods, however, are sweet and are fed to livestock when other feed is in short supply. An extract of inner bark tissues is used for treating diarrhea. MASILUNGAN *et al.* (1958) tested leaves of the acacia tree and found certain extracts inhibitory to *S. aureus, S. lutea, B. subtilis, E. coli,* and *M. tuberculosis.* Water extracts generally were more active than were those obtained with organic solvents.

FISCHER (1952) extracted a phagicidal and virucidal compound from fruits of the Norway maple. The substance, named "acerin", destroyed the vaccinia virus and the bacteriophage able to infect cells of *E. coli.* Destruction occurred even in the presence of the host cells.

WINTER and WILLEKE (1952) tested a variety of grasses, both green and dried, for inhibitory activity against *B. subtilis, E. coli,* and *S. aureus.* Generally extracts of dried leaves were inhibitory and of green leaves were stimulatory. Grasses tested include *Koeleria alpicola* and *pyramidata, Hordeum murinum* (wild barley), *Agropyrum caninum* (dog-quack grass) and *repens* (quack grass), *Poa chaixi, nemoralis* (meadow grass), *palustris,* and *pratensis* (Kentucky blue grass), *Festuca gigantea* (fescue), *glauca* (blue fescue), and *ovina* (sheep's fescue), *Phleum nodosum, phleoides,* and *pratensis* (timothy).

The unsaponifiable fraction of cod-liver oil, according to EMMERIE *et al.* (1950), contains a bacteriostatic agent especially effective against *M. tuberculosis.* This fraction, a viscous oil, did not contain sterols, vitamin A, or vitamin D. Inhibition was noted at concentrations of 1.0 to 3.0 µg./ml. Hydrogenation of the oil reduced its bacteriostatic activity.

Body tissues have been found to possess germicidal activity. BLOOM and PRIGMORE (1952) indicate that a substance able to inhibit *B. anthracis* has been recovered from the cecum, thymus, and pancreas of various animal species. A basic polypeptide with a high content of lysine was believed to be the active material. *E. coli*, *S. aureus*, β-hemolytic streptococci, and *B. megaterium* were inhibited by thyroid tissue, and *B. subtilis* by beef spleen, thymus, and pancreas tissue.

III. Antibiotics produced by organisms employed in the manufacture of some foods

Many commonly consumed foods are produced by means of certain microbial fermentations. The lactic acid fermentation is, perhaps, the most frequently used since production of many foods including most cheeses, cultured milks, sauerkraut, cucumber pickles, green olives, etc. is dependent on it. Molds are used in the ripening of some cheeses and formation of acetic acid is needed to produce vinegar. Development of propionic acid is desired in certain cheeses.

Many of the organisms employed in these fermentations can produce antibiotic-like substances in addition to the other changes brought about in the substrate. This section summarizes the most significant information on production of the inhibitory substances, susceptibility of other organisms to these materials, and potential uses of the antimicrobial agents. Antibiotic production has been associated with the lactic streptococci, lactobacilli, propionibacteria, acetobacters, and some other microorganisms commonly associated with food fermentations.

a) Homofermentative lactic streptococci

Lactic streptococci able to produce only very minute amounts of substances other than lactic acid from carbohydrates are designated as homofermentative. Species included in this definition are *S. lactis* and *cremoris*. Inhibitory activity has been associated with *S. lactis* at least since 1928 when ROGERS (1928) reported it was able to inhibit *Lactobacillus bulgaricus* and with *S. cremoris* since 1933 when WHITEHEAD (1933) found it produced polypeptides which retarded growth of other lactic streptococci. Later the inhibitory substance produced by *S. cremoris* was designated "diplococcin" and the one recovered from *S. lactis* was called "nisin".

Diplococcin was described by OXFORD (1944) who observed that the substance inhibited certain strains of *S. cremoris* and *lactis*, and *S. aureus*. The antibiotic was found to be water-soluble but insoluble in absolute ethanol. It was thought to be a protein-like substance devoid of sulfur or phosphorus. The antibiotic was stable at pH 4.0 and 100° C. but not at pH 6.0 to 7.0. A yield of 20 mg. of antibiotic/g. of dried cells was obtained.

Although the purified diplococcin never appears to have left the realm of a laboratory curiosity, this situation is not true of the antibiotic produced by *S. lactis*. MATTICK and HIRSCH (1944) described a substance recovered from *S. lactis* which inhibited bacteria in the genera *Bacillus*,

Clostridium, and *Lactobacillus;* staphylococci were found less susceptible and gram-negative bacteria were unaffected. As discussed below, the substance was later designated nisin. In some instances (HOYL and NICHOLS 1948) production of these inhibitory substances may be detrimental if mixed strain starter cultures are being maintained. The presence of antibiotic-producing strains may retard acid formation by the culture and these strains may become dominant after a few transfers.

1. The nature of nisin. — BERRIDGE *et al.* (1952) reported that nisin consisted of at least four active polypeptides designated by them as *A, B, C,* and *D.* Fractions *A* and *B* were similar in their activity against *S. agalactiae* but C and *D* were only $^{1}/_{5}$ as potent as the first two components. Nisins *A, B,* and *C* were found to contain leucine and/or isoleucine, valine, alanine, glycine, proline, aspartic acid, histidine, lysine, methionine, lanthionine, and cystathionine or allocystathionine. Nisin *D* differed from *A, B,* and *C* in that it contained glutamic acid but no valine or methionine. These authors suggested that nisin, in some ways, resembled the antibiotic subtilin. Later NEWTON *et al.* (1953) and CHEESEMAN and BERRIDGE (1959) showed nisin *A* to contain the amino acids leucine and/or isoleucine, methionine, valine, proline, alanine, glycine, serine, aspartic acid, histidine, lysine, β-methyllanthionine, and lanthionine. The leucine-isoleucine combination, lysine, and β-lanthionine were present in highest concentrations. They also concluded that nisin-*A* had a molecular weight of 7,000 and contained no amino or carboxyl end groups except the ε-group of lysine and the imino group of histidine.

The solubility of nisin is dependent on the pH of the solvent (HAWLEY 1957 a, WODSAK 1962). At pH 7.0 in water the solubility is approximately 75 µg./ml. This increases rapidly to 1000 µg./ml. at pH 5.6 and to 12,000 µg./ml. at pH 4.2. The presence of protein in the solvent appears to increase solubility probably through adsorption. Crystalline nisin and high-titer preparations are most readily soluble in a 0.02 *M* hydrochloric acid solution.

In the dry state nisin may be stored for years at room temperature with no apparent loss of activity. In solution, however, stability is dependent on pH, temperature, and nature of the substrate (HAWLEY 1957 a). Rapid inactivation occurs in an alkaline solution although, in some instances, this is reversible. At 100° C. complete inactivation occurs in two hours at pH 8.0, in one hour at pH 9.0, and in 30 minutes at pH 10.0. At pH 6.0 activity is reduced 25 percent by a one-hour exposure to 100° C. The antibiotic is partially inactivated by salivary ptyalin, is digested by trypsin, and is completely inactivated by irradiation (2.7 megarads) (HAWLEY 1958, LEY and HALL 1960). When fed to pigs and chickens, nisin failed to stimulate growth (BARBER *et al.* 1952).

2. Susceptibility of organisms to nisin. — MATTICK and HIRSCH (1947), shortly after their initial reports on nisin, found the antibiotic at concentrations of 0.25 to 500 units/ml. able to inhibit a variety of streptococci, staphylococci, *M. lysodeikticus,* pneumococci, neisseria, *Bacillus* spp., clostridia, corynebacteria, mycobacteria, lactobacilli, actinomyces, and *Erysi-*

pelothrix spp. Activity of nisin persisted in media containing serum, blood, or milk. Gram-negative organisms were found to be resistant to the antibiotic.

Inhibition of clostridia by nisin appears to have been of greatest interest to research workers. Undoubtedly this stems from the importance of some bacteria of this type in the spoilage of cheese. HIRSCH *et al.* (1951) recovered an anaerobe from Swiss cheese which was inhibited in milk by a nisin-forming culture of *S. lactis.* LIND (1958), using artificial media inoculated with low concentrations of different clostridia, found some were inhibited while others were unaffected by nisin. Other experiments by GARCIA (1959) also were conducted in artificial media and suggested that nisin inhibited reproduction of *Clostridium butylicum* but had no effect on its metabolism of carbohydrate. An acid pH and addition of sodium phosphate or sodium citrate to the medium enhanced the action of nisin while the presence of hydrogen peroxide and catalase was without influence.

SJÖSTRÖM and WILLART (1953) found nisin more active against lactose-fermenting than lactate-fermenting clostridia although the latter type was somewhat sensitive. LIPINSKA (1956 a and b) demonstrated that growth of clostridia in Edam cheese was prevented by the use of lactic streptococci able to produce antibiotics. GALESLOOT (1957 b) reported that the presence of nisin in processed cheese extended its shelf-life because butyric acid organisms such as *Clostridium tyrobutyricum, sporogenes,* and *tetanomorphum* were inhibited. Inhibition of *C. tyrobutyricum* was enhanced by treating antibiotic-producing strains of *S. lactis* with ultraviolet light according to results of CSISZAR and PULAY (1956).

The effect of nisin on some clostridia was examined in greater detail by HIRSCH and GRINSTED (1954). They found that spores of *C. butyricum* were more susceptible to the antibiotic than were vegetative cells. Cells and spores of *C. sporogenes* and *bifermentans* were more resistant to nisin than those of *C. butyricum.* Additional tests suggested the antibiotic was actually sporicidal in one instance *(C. butyricum)* but not in others.

RAMSEIER (1960 and 1961) extensively studied the effect of nisin on *C. butyricum.* He observed that a) nisin was bactericidal against young and old vegetative cells of the organism and resistance to nisin was dependent on the age of the cells and concentration of the antibiotic, b) the antibiotic was neither sporostatic nor sporicidial but acted on germinated spores, c) the bactericidal effect of nisin depended on the size and age of the inoculum, pH of the medium, and composition of the medium, d) nisin was neutralized by soap while divalent cations reversed the bacteriolytic action of the antibiotic, and e) nisin acts in a manner similar to cationic detergents and the surface-active polypeptide antibiotics.

Nisin was found to be active against *Clostridium perfringens (C. welchii)* by EASTOE and LONG (1959). They noted two Reading units of antibiotic per ml. were needed to inhibit growth of vegetative cells while a concentration of 40 units/ml. was required to prevent colony formation from spores. Inhibition of *C. septicum* was reported by MATTICK and HIRSCH (1949) who also noted activity of the antibiotic against those streptococci able to cause mastitis in dairy cattle.

Although most experiments on susceptibility of organisms to nisin have centered around the clostridia, some investigators have been concerned with other bacteria. Results of their tests will be considered below. GALESLOOT (1957 a and b) found nisin able to inhibit *Streptococcus diacetilactis, Leuconostoc citrovorum, L. plantarum, Lactobacillus bifermentans,* and propionibacteria. Some strains of *S. lactis* and *cremoris, L. citrovorum,* and *L. plantarum* were found able to destroy nisin while coliforms were unaffected by the antibiotic and did not destroy it. Some strains of *Streptococcus thermophilus* were found to be inhibited by nisin in tests of SHAHANI (1962), although cultures of *S. lactis, L. bulgaricus,* and *B. subtilis* and *cereus* were unaffected. Certain nonpathogenic strains of *S. aureus* were found sensitive to the antibiotic. HALASZ and JUHASZ (1963) verified results SHAHANI (1962) obtained with *B. subtilis.* They did note, however, that the presence of nisin reduced the heat or x-ray exposure needed to destroy *B. subtilis.* Results with *B. subtilis* directly opposite to those just considered were reported by MOYCHO and GROMSKA (1956) who found this organism and *P. fluorescens* were inhibited by nisin in milk media. GARVIE and STONE (1953), while studying the effect of nisin on *B. cereus,* came to the same conclusion as did SHAHANI (1962). Antibiotic-producing strains of *S. lactis* were isolated by LUKOVNIKOVA (1961) who found the antibiotic able to inhibit *S. diacetilactis, E. coli,* and *B. cereus* var. *mycoides.*

The sensitivity to nisin of *B. coagulans,* an organism often associated with the spoilage of heat-processed canned foods, was investigated by CAMPBELL and SNIFF (1959). They found all of 31 strains were inhibited by 5.0 µg. of nisin/ml. while 90 percent of the strains in basamin broth and 61 percent of the strains in tomato juice failed to grow in the presence of 1.0 µg. of antibiotic/ml. Both vegetative cells and spores were found sensitive to nisin. The authors suggested nisin combined with a mild heat treatment might be used to process tomato juice. Heat resistance of spores of *Bacillus stearothermophilus,* according to THORPE (1960), was not affected by nisin although germination and outgrowth may be.

In vivo experiments with nisin by GOWANS *et al.* (1952) and HIRSCH and MATTICK (1949) revealed a) an LD_{50} of 30 mg./kg. with intravenous injection and 200 mg./kg. with intramuscular injection, b) solubility in blood of 5,000 to 20,000 Reading units/ml., c) reduction of tuberculosis infections in guinea pigs injected with the antibiotic, and d) effectiveness against experimental infections in mice caused by *S. pyogenes* and *S. aureus.* More recent tests by FRAZER *et al.* (1962) demonstrated that dosages of nisin up to 1,400 times the normal human intake had no detrimental effect on male rats. CARLSON and BAUER (1957) demonstrated that resistant strains of *S. aureus* developed when the organism was grown in the presence of sublethal nisin concentrations. A substance acting on nisin in a fashion similar to penicillinase on penicillin was recovered from nisin-resistant strains of bacteria. Their tests indicated further that viruses were unaffected by nisin. HIRSCH (1950) found resistant strains of *S. agalactiae* developed when the organism was grown in the presence of sublethal concentrations of nisin. More recently LIPINSKA and STRZALKOWSKA (1959) were able to develop strains of *S. lactis, cremoris,* and *thermophilus, Leuconostoc dex-*

tranicum, and *L. helveticus* which were resistant to 17 to 100 units of nisin/ml. of milk.

3. Destruction of nisin by microorganisms. — Observations of some workers already reported (GALESLOOT 1957 b, CARLSON and BAUER 1957) have suggested that certain microorganisms can destroy nisin. Additional information on this subject is summarized below.

KOOY (1952) and PETTE and KOOY (1953) reported that some strains of *L. plantarum* present in raw milk could destroy nisin. They noted further that inoculation of milk with *L. casei* prevented inactivation of the antibiotic even in the presence of nisin-destroying strains of *L. plantarum.* Fifty percent of the streptococci found in raw milk, according to GALESLOOT (1956 b) were also able to destroy the antibiotic. RAMSEIER (1959) found *E. coli* and *A. aerogenes* unable to inactivate the antibiotic.

Although, according to CHEESEMAN and BERRIDGE (1954), nisin was not affected by penicillinase, this antibiotic was found to be inactivated by nisinase in tests of ALIFAX and CHEVALIER (1962). Nisinase was obtained from some strains of *S. thermophilus* and was found to be most active at pH 7.0.

4. Measurement of nisin. — Methods to detect and measure the concentration of nisin in a particular substrate include both tube-dilution and agar-plate procedures. Initially it was observed by MATTICK and HIRSCH (1947) that nisin failed to diffuse through agar at pH 7.0. Hence they suggested use of a tube test employing a hemolytic strain of *S. agalactiae.* A sterile broth containing glucose, peptone, and sodium chloride was inoculated with sufficient of a 24-hour old culture to provide 2,000 to 10,000 viable organisms/ml. The inoculated broth was then used to prepare serial dilutions of the test substance, the dilutions usually being made in powers of two. Results were read when control tubes showed growth, usually after 16 to 20 hours at 37° C. The titer of nisin was expressed as the dilution of the last tube in which growth failed to occur. HIRSCH (1950) reported a tube test based on changes in the pH of the medium brought about if the nisin concentration was low enough to permit the eventual growth of *S. agalactiae.*

BERRIDGE and BARRETT (1951 and 1952) developed a tube dilution method for assay based on turbidity measurements made on cultures of *S. agalactiae* growing in broth fortified with varying concentrations of the substance thought to contain nisin. They were able to obtain results after 30 minutes of incubation. This technique permitted detection of 2.9 to 7.2 units/tube containing 17.5 ml. of liquid.

CSISZAR and PULAY (1956) employed *C. tyrobutyricum* in a tube assay. Tubes of liver broth and the test substance were inoculated with the culture, refrigerated for two hours to permit equilibration, then incubated at 37° C. After four hours, control tubes were checked at 30-minute intervals for gas formation. When gas appeared, tubes containing the antibiotic were incubated an additional 30 minutes, checked for gas production, and the titer expressed as the highest dilution which suppressed gas formation.

The use of a thermophilic bacterium *(Thermobacillus tarbellicus)* and incubation at 68° C. has been suggested by RAIBAUD (1956) for the rapid assay of nisin. Later RAIBAUD (1957) suggested that a complex was formed by nisin and a metabolite required by the organism. More recently STUMBO *et al.* (1964) employed "heat-damaged" spores of *B. stearothermophilus* in a tube test to assay for nisin in foods. The test is based on inhibition of colony development in agar contained within the tube after incubation at 130° F. for 24 hours. The authors claimed this technique detected as little as 0.3 unit (0.0075 p.p.m.) of nisin in products such as skim-milk, chocolate-flavored milk, cream-style corn, peas, mushroom soup, lima beans, beets, okra, tomato juice, peaches, apples, and sour cherries.

A variety of *L. helveticus* was also found suitable for detection of nisin by means of the tube dilution test (TEPLY 1961). This organism, claimed able to detect 0.8 unit/ml. provided results in 5.5 to 7.0 hours.

The use of a sensitive lactic streptococcus (e. g., *S. cremoris*) to estimate the concentration of nisin in a test substance has been suggested by several authors. GALESLOOT and PETTE (1956) used litmus milk to which was added varying amounts of the test material together with a constant quantity of a 24-hour old culture of *S. cremoris*. Tubes were then incubated at 30° C. for 24 hours and the smallest volume of test substance able to inhibit the culture was noted. If a nisin standard is available duplicate tubes can be prepared and by comparison an estimate of nisin concentration in the unknown sample can be obtained. These authors found the procedure described above suitable for use with nisin-producing cultures of *S. lactis* as well as cheese to be assayed for its nisin content.

CHEVALIER and MOCQUOT (1959) used *S. cremoris* and milk fortified with rennet to estimate the concentration of nisin in test materials. Production of acid by *S. cremoris* hastens coagulation of milk by rennet and so any delay in coagulation could be attributed to a reduction in acid development by the streptococcus and would reflect the concentration of nisin present. The conventional resazurin test was modified by FRIEDMANN and EPSTEIN (1951) for use in estimating the concentration of nisin in a test material. Milk containing appropriate dilutions of the test material was inoculated with *S. cremoris*, incubated at 30° C. for 15 minutes, fortified with dye, and reincubated for five minutes. A lack of color change is associated with destruction of the organism.

Several agar-plate methods have been found suitable in the assay of nisin. BEACH (1952) found the addition of a non-ionic surface active agent to the agar medium enhanced diffusion of nisin so that appreciable zones of inhibition were produced with *S. agalactiae*. Use of *M. phlei* in an agar-plate assay was suggested by GOWANS *et al.* (1952). They refrigerated inoculated plates for 24 hours to permit diffusion of the nisin, followed by a 24-hour incubation at 37° C. and examination for zones of inhibition. MOCQUOT and LEFEBVRE (1956) found an agar-plate method suitable for detection of nisin in cheese. A cylinder of cheese cut from a plug was placed in a petri dish and covered with 20 ml. of an agar medium previously inoculated with *L. lactis* or *S. cremoris*. When *L. lactis* was used, zones of inhibition could be detected in four hours while *S. cremoris* required an overnight incubation. Larger zones of inhibition were obtained when the plate was refrigerated for 20 hours prior to incubation.

Recently TRAMER and FOWLER (1964) suggested two different agar-plate techniques for estimation of nisin in foods. The first employs an agar medium fortified with Tween 20 (polyoxyethylene sorbitan monolaurate) and *M. flavus* as the assay organism. Circular holes cut into the solidified, inoculated agar in a petri plate are filled with test solutions after which the plate is incubated overnight at 30° C. and then inspected for zones of inhibition. This procedure detects down to 0.5 unit of nisin/ml. The second procedure is a reverse-phase disc assay using heat-shocked spores of *B. stearothermophilus* and a non-nutrient saline agar. Specially prepared nutrient discs are dipped into the test solution, placed on previously inoculated and solidified agar within a petri plate, and incubated at 55° C. for 18—24 hours. Zones of inhibition were produced by 0.5 to 1.0 unit of nisin/ml.

5. Uses of nisin. — Nisin has been suggested for use in natural and processed cheeses to control, principally, the growth of anaerobic spore-forming bacteria. Patents have been issued (MATTICK and HIRSCH 1954 and 1956) for such use. In their patents the authors discuss its suitability for use in the manufacture of Gruyère, Comte, Beaufort, Emmenthal (Swiss), Brie, Danish-Swiss, and Swedish Manor House cheeses as well as melted (e. g., processed) cheeses. Many experiments have been performed on the use of nisin in cheeses. Results of some will be summarized below.

Beneficial effects from the use of nisin were reported by KOOY and PETTE (1952 a and b). They found the antibiotic able to inhibit gas formation by *C. tyrobutyricum* in cheese, provided *L. plantarum* did not inactive nisin.

BYLUND *et al.* (1954) used antibiotic-producing strains of *S. lactis* in the manufacture of Herrgard cheese. Even when cheese milk was inoculated with anaerobic butyric acid bacteria, defects failed to appear in cheese made with the inhibitory *S. lactis* strains. ROSELL and MATALLANA (1956) were able to control defects in cheese made in a factory heavily contaminated with *C. tyrobutyricum* when they used nisin-producing strains of *S. lactis* as the starter culture. Inhibition of butyric acid bacteria in Edam cheese made with a nisin-producing starter was observed by LIPINSKA (1956 b). Later LIPINSKA (1961) noted that use of nisin-resistant starters in making Edam cheese from milk fortified with nisin resulted in a more rapid depletion of lactose and an improved flavor. GALESLOOT and PETTE (1957) found that eye formation in Edam cheese made with a nisin-producing starter was unsatisfactory, especially when pasteurized milk was used. Later GALESLOOT and HASSING (1964) established that the nisin content of Edam and Gouda cheeses was practically stable during ripening, provided the cheese was made from pasteurized milk. In other tests, GALESLOOT *et al.* (1957) noted good Edam cheese could be made using nisin-producing starter cultures although such cultures generally retarded ripening of the cheese. STÜBER and BINDER (1959 a and b) successfully used nisin-producing starters to manufacture Edam and Gouda cheeses from "silo" milk containing large numbers of butyric acid bacteria. LIPINSKA *et al.* (1962) verified the successful production of Edam cheese with nisin-forming starter streptococci.

Experiments on the use of nisin in Swiss cheese manufacture (RITTER 1957, WINKLER and FRÖHLICH 1957 and 1958) indicate the antibiotic is unsuitable in this instance. Abnormal ripening of the cheese was generally observed and this was attributed to the inhibitory action of the nisin on lactobacilli and propionibacteria — normal ripening agents of Swiss cheese. PULAY (1956) employed four kinds of cheese. He concluded that using nisin-producing starters in their production only delayed (but did not prevent) gas formation. It was thought butyric acid bacteria regain activity if the antibiotic is exhausted. The author also suggests that presence of nisin in cheese may encourage gas production by coliform bacteria.

Use of nisin to control the growth of clostridia in processed cheese has been suggested by MCCLINTOCK *et al.* (1952), BERRIDGE (1953), HAWLEY (1957 a and b), LILLEY (1958), and WODSAK (1962). LIPINSKA and STRZALKOWSKA (1958) added nisin preparations to natural cheese with symptoms of a butyric acid fermentation when it was converted to processed cheese. Less than 50 percent of nisin-treated cheese spoiled while 100 percent of untreated cheese underwent the butyric acid fermentation. BARTUSKOVA-CERNIKOVA (1962) found less nisin was required to suppress gas formation in processed cheese than in milk. Use of nisin in processed cheeses retarded gas formation and prevented white rot according to GORNER and SIMKOVICOVA (1961). TSUGO *et al.* (1962) found 50 units of nisin/g. of cheese sufficient to prevent gas formation in processed cheeses made from Swiss and Cheddar cheeses while GORNER and KABATOVA (1963) reported 100 to 150 units/g. failed to suppress microbial changes in a processed cheese having a heavy localized microbial infection brought about by adding contaminated caraway seed.

Nisin has been suggested for use in the preservation of foods other than cheese. Included are canned foods, vegetables, fruit, meats, fish, soups, evaporated milk, chocolate milk, sterile whole milk, cream, and butter. HAWEY (1957 a) reports that nisin inhibits development of flat sour, thermophilic anaerobic, and putrefactive anaerobic bacteria in canned foods. Of all microorganisms those just mentioned are most likely to survive canning heat processes. HAWLEY (1957 b) does point out that nisin does not inhibit all food spoilage organisms and cannot be considered as a general food preservative. Nisin has been found beneficial to minimize souring of hams by lactobacilli and to prevent development of staphylococci. Foods contaminated with a variety of organisms will often contain some with resistance to nisin. Recently HEINEMANN *et al.* (1964) demonstrated that the use of 80 units of nisin/ml. and a heat treatment equivalent to an *F* value of 3.0 (heat at the can center in terms of minutes at 250° F., assuming instantaneous heating and cooling) resulted in a beverage-quality, commercially sterile chocolate-flavored milk. Use of nisin and a minimum heat exposure to process evaporated milk reduced the loss of vitamins but had no effect on the nutritive value of the protein when compared to that of conventionally processed evaporated milk (GREGORY *et al.* 1964). In another recent report, WHEATON and HAYS (1964) observed that nisin added to cream style corn and chow mein at a concentration of 1.0 p.p.m. prevented flat sour spoilage even in inoculated packs. Although when added at a concentration of 2.5 p.p.m. it failed completely to prevent spoilage by putrefactive anaerobes, the incidence was reduced. VAS (1963) found nisin helpful in the preservation of tomato juice and a tomato juice-yellow paprika product. The heat treatment used in conjunction with nisin could be reduced.

GIBBS and HURST (1964) conducted tests to determine if nisin could be used in conjunction with reduced heat treatments to preserve soups. Cream of chicken and lentil soups were prepared and nisin, even when added at concentrations of 500 units/g., failed to prevent all spoilage in the canned products autoclaved for 20 minutes (including come-up time) at 120° C.; 75 percent of nisin activity was destroyed during autoclaving. Use of nisin to preserve bacon was also without success for most of the antibiotic was lost during curing and subsequent storage. Nisin-resistant lactobacilli and micrococci appeared on the treated bacon.

Since nisin is active against *S. agalactiae,* the causative agent of some forms of bovine mastitis, it could be expected that someone might use the antibiotic to eliminate the mammary gland infection. TAYLOR *et al.* (1949) found a single intramammary infusion of five-million units/quarter cured 35 of 37 cases while use of 2.5 million units each freed 30 of 35 quarters from infection with *S. agalactiae.* Similar results were obtained when *S. aureus* caused the mastitis but not when the etiological agent was *S. dysagalactiae.*

6. Production of nisin. — It has been indicated previously that nisin formation is associated with *S. lactis.* HIRSCH and WHEATER (1951) examined 510 pure cultures of streptococci recovered from raw and pasteurized milk, cheese, human, bovine, and pig feces, and human nose and throat swabs. They failed to find streptococci other than *S. lactis* able to produce inhibitory substances. Simultaneously HIRSCH and GRINSTED (1951) suggested that the antibiotics produced by *S. lactis* and *cremoris* differed sufficiently to permit use of this characteristic as a means for differentiating between the organisms.

Nisin can be developed in a food material such as cheese if an appropriate fermentation takes place during its production (HAWLEY and HALL 1957). If the antibiotic is to be used in another food it must be produced separately and processed in order to make it available. BERRIDGE (1949) described a method for the production of nisin which included a) growth of *S. lactis* in a glucose-yeast extract broth during which time the nisin was produced, b) concentration of nisin by adsorption with a chloroform emulsion, c) fractionation to increase activity in phosphate buffer, d) further purification by fractional precipitation with sodium chloride, and e) crystallization of the nisin. Final treatment involved recrystallization to form needles from a cooled ethanolic solution at pH 4.2.

HIRSCH (1951) found that stable mutant colonies picked from a parent strain of *S. lactis* could be used to improve the yield of nisin. Colchicine and camphor treatments did not enhance production of the antibiotic. Maximum yields of nisin were obtained when, during fermentation, there was present sufficient glucose and pantothenate, an incubation of 23 to 30° C. was used, and the pH was maintained at 6.0. GALESLOOT (1957 b and 1959) found that non-nisin producing mutants can arise in a nisin-producing culture thus reducing the yield of the antibiotic. Furthermore, the cultures were very sensitive to attack by bacteriophages. RITTER (1963) noted that nisin production could be restored if faulty cultures were grown together with a resistant organism such as *L. helveticus.* A patent was issued to FALCONER (1952) for production of nisin according to procedures embodying many of the suggestions made by BERRIDGE (1949) and HIRSCH (1951). FALCONER (1952) suggested addition of 0.02 to 1.0 percent of sulfate to the fermentation medium and recovered the antibiotic with an acidified acetone solution or a four-carbon aliphatic alcohol adjusted to a pH between 0.5 and 3.5. A revised procedure claimed to minimize losses and employing salting out, solvent extraction, and fractional precipitation with acetone was described by CHEESEMAN and BERRIDGE (1957). HAWLEY (1955 and 1957 c) described a method for the production of nisin in which skim-milk is cultured with an appropriate strain of *S. lactis,* the pH is then adjusted to 3.0, the cultured milk is heated to 90° C., curd is removed, and whey (which contains the nisin) is concentrated and eventually dried so that a powdered nisin preparation results. The process was modified further by HAWLEY and HALL (1960 a and b) who used a foam-technique for recovering the nisin. Another improvement suggested by HALL (1963) consisted of pre-treating milk with proteolytic enzymes. WINKLER and MATHA (1956) compared different media for their suitability in nisin production and obtained highest yields using milk fortified with glucose and an incubation period of four days.

b) Heterofermentative lactic streptococci

Lactic acid bacteria able to produce a variety of compounds in addition to lactic acid are designated as heterofermentative. The most common of these are *L. citrovorum* and *dextranicum* and *S. diacetilactics.* These bacteria are commonly used in certain dairy fermentations to produce desired flavors and aromas. They also appear incidentally in fermentations of the type employed to produce sauerkraut and cucumber pickles. When compared to the homofermentative lactic streptococci, relatively little attention has been devoted to antibiotic production by these organisms.

RITTER (1945) described experiments in which betacocci (*Leuconostoc* spp.) inhibited two strains of *S. lactis.* More recently MATHER and BABEL (1959) prepared a cottage cheese creaming mix composed in part of skim-milk cultured with *L. citrovorum;* use of this mix was associated with inhibition of *Pseudomonas fragi* and *putrefaciens,* and coliform bacteria. MARTH and HUSSONG (1963) found that *L. citrovorum,* when grown in skim-milk and allowed to ferment citric acid, produced a substance able to inhibit *S. aureus, A. aerogenes, Alcaligenes viscolactis, E. coli,* and *P. fragi* and *fluorescens.* Differences in their ability to produce the inhibitory substance was found to exist between various strains of *L. citrovorum.* Activity of the filtrates was related to pH, being highest at 4.5 and substantially less at 5.1. MARTH and HUSSONG (1962) also diluted filtrates to the level that cultured material might be present in cottage cheese if a mix as described by MATHER and BABEL (1959) were used; under such conditions, of all the test organisms only one of four strains of *P. fluorescens* and one of two strains of *P. fragi* were inhibited.

S. diacetilactis differs from the other heterofermentative lactic streptococci in that it produces a higher concentration of lactic acid. This organism is often found in mixed starter cultures used in the dairy industry. GIBSHMAN and BELOUSOVA (1958) observed that *S. diacetilactis* produced substances inhibitory to *S. lactis* and *E. coli*. COLLINS (1961) studied six strains of *S. diacetilactis*. Some strains were able to retard growth of *S. cremoris*, *S. lactis* and other strains of *S. diacetilactis*. One strain was able to inhibit *S. lactis* only. The antibiotic produced by *S. diacetilactis* behaved similar to one formed by *S. cremoris* and dissimilar to that of *S. lactis*.

Diacetyl is the principal aroma and flavor compound produced by heterofermentative lactic streptococci in their metabolism of citrate and/or a sugar. This compound, according to HEDGECOCK and JONES (1950), does have some antimicrobial activity when present at concentrations of 0.05 to 0.25 percent. These levels are appreciably higher than is the concentration of diacetyl normally associated with the metabolism of the heterofermentative streptococci. Hence it is reasonable to assume that although diacetyl may contribute to total antimicrobial activity, other substances are produced which possess the bulk of the activity.

c) Lactobacilli

Bacteria in the genus *Lactobacillus* are employed in the manufacture of fermented milks such as yoghurt, acidophilus milk, Bulgarian buttermilk, and others. Some cheeses, such as Swiss and Parmesan, are produced with the aid of acid formed by certain lactobacilli whereas others, like Cheddar, are dependent (in part) on the proteolytic and lipolytic action of lactobacilli for development of flavor during ripening. Elaboration of antibiotic substances has been associated with many of the lactobacilli involved in the production of these and other milk products. The individual organisms and the finished products which they help to produce are considered below.

1. Lactobacillus acidophilus. — Numerous authors have associated the production of an antibiotic substance with this organism. NERI (1961) observed that two strains of *L. acidophilus* inhibited growth of *B. subtilis*, *S. aureus*, *E. coli*, and *S. typhi* and *dysteneriae*. Another culture of *L. acidophilus*, however, failed to inhibit the test cultures. Inhibition of *M. tuberculosis* by *L. acidophilus* (especially when grown in whey) was reported by TOMIC-KAROVIC *et al.* (1959). WINKLER (1953) found milk cultures of *L. acidophilus* able to inhibit eight species of bacteria in the genus *Propionibacterium*. Included was *P. shermanii*, the organism responsible for flavor and eye formation in Swiss cheese. The inhibition of *P. fluorescens* by *L. acidophilus* was noted by POLONSKAYA (1952).

Cultures of *L. acidophilus* obtained from mice, rats, rabbits, hamsters, and man were found to produce an antimicrobial substance in experiments by VINCENT *et al.* (1959). This substance was designated "lactocidin" and was measured in terms of units, one unit being defined as that amount of active material needed to limit the growth of *S. faecalis* to half the value obtained in control tubes maintained under similar conditions. Lactocidin at concentrations of 1.0 to 6.0 units/ml. of tryptose broth, pH 4.7, inhibited the following organisms: *P. vulgaris, S. enteritidis. Pasteurella multocida, P. aeruginosa,* a paracolobactrum, *E. coli, M. smegmatis* and *phlei, Tricophyton gypseum, Microsporum lanosum, S. aureus, Corynebacterium xerosis, Gaffkya tetragena, Staphylococcus epidermidis, B. subtilis, S. faecalis, P. notatum,* and other strains of *L. acidophilus.* When a broth-serum medium at pH 7.0 was used only *P. vulgaris, S. enteritidis, P. multocida, Neisseria catarrhalis, E. coli, P. aeruginosa,* a paracolobactrum, *M. smegmatis,* and *S. aureus* and *epidermidis* were inhibited. DeKlerk and Coetzee (1961) observed that the antibiotic produced by some strains of *L. acidophilus* was able to inhibit lactobacilli including other strains of *L. acidophilus, bulgaricus, salivarius, helveticus,* and *fermenti.*

This antibiotic substance, also designated "acidophilin", was reported heat stable and most toxic to *E. coli* when in a pH range of 5.0 to 5.6 (Polonskaya 1952). Dubinskii (1958), however, noted the inhibitory agent of *L. acidophilus* was unstable and that freshly prepared fermented milks were the best sources.

Koroleva (1958) found antibiotic production by this lactobacillus was best when skim-milk cultures were stored at 3.0° to 5.0° C. and transferred at 1.0- to 1.5-month intervals. Use of frozen cultures stored at —25° C. and transferred at six-to-seven month intervals was equally satisfactory.

A method for the detection of antibiotic activity in cultures of *L. acidophilus* was proposed by Sabine (1963). The procedure, a modification of one suggested earlier by Ritter (1956 b), employed a petri plate containing an agar medium inoculated with *S. aureus.* Round holes were cut into the solidified agar and the agar thus separated from the main body was removed. Agar seeded with *L. acidophilus* was introduced into the open areas and allowed to solidify. After an appropriate incubation period active cultures produced zones in which the test organism was inhibited. Inhibition did not result from acid production by *L. acidophilus* but was caused by an inhibitory substance produced by the cells.

2. Lactobacillus helveticus. — Winkler (1953) noted that *L. helveticus,* one of the acid-forming bacteria sometimes used as a starter culture in the manufacture of Swiss cheese, produced a substance able to inhibit bacteria in the genus *Propionibacterium.* Included was *P. shermanii,* the organism commonly associated with flavor and eye formation in Swiss cheese. Whey cultures of *L. helveticus,* according to Ritter (1956 b), were able to inhibit *E. coli* and *A. aerogenes.* Inhibition was attributed to a substance other than lactic acid. Cultures and culture filtrates of *L. helveticus* were examined by Meewes and Milosevic (1962) for their effect on gas formation by *E. coli.* Both live cells and culture filtrates were able to reduce gas production by *E. coli.* Forty-eight hour old cultures and filtrates prepared from such cultures were more active than similar materials only 24 hours old.

Stocker (1954) reported that *L. helveticus* failed to prevent growth of and gas formation in cheese by *Saccharomyces fragilis,* while Neri (1961) observed that *B. subtilis, E. coli, S. aureus,* and *S. typhi,* and *dysenteriae* were not inhibited by two strains of *L. helveticus.*

One strain of *L. helveticus* able to inhibit *S. aureus* was isolated from French Gruyère cheese by WHEATER *et al.* (1951). The antimicrobial substance was 20 times more powerful than nisin but was very heat labile.

3. Lactobacillus lactis. — An antibiotic substance able to inhibit *E. coli* and *A. aerogenes* was associated with *L. lactis* by RITTER (1956 b). It was similar in its action to that produced by *L. helveticus* but appeared at higher concentrations in *L. lactis* cultures. RITTER (1956 b) noted further that the antibiotic activity was not related to formation of hydrogen peroxide by *L. lactis* or *helveticus.* Neutralization of the peroxide with catalase left the antibiotic with its activity undiminished.

L. lactis, according to WINKLER (1953), inhibited a few propionibacteria but *P. shermanii* was not among them. Results reported by STOCKER (1954) suggest that *L. lactis* had an inhibitory effect on growth of, and gas production by, *S. fragilis* in milk and cheese.

4. Lactobacillus bulgaricus. — STOCKER (1954) found that *L. bulgaricus* tended to reduce growth of and gas formation by *S. fragilis* in milk and cheese. DUBINSKII (1958) found *L. bulgaricus* produced its unstable antibiotic at the beginning of the logarithmic phase while NERI (1961) reported the inhibitory substance failed to retard growth of *B. subtilis, E. coli, S. aureus,* and *S. typhi,* and *dysenteriae.*

5. Other lactobacilli. — An antibiotic, "lactolin", was isolated from *L. plantarum* by KODAMA (1952). Cultures of *B. subtilis, E. coli, S. aureus,* and *S. typhi,* and *dysenteriae* were inhibited by some strains of *L. casei* but not by *L. leichmannii* or other strains of *L. casei* according to results of NERI (1961). HIRSCH *et al.* (1952) recovered 55 cultures of lactobacilli from young Gruyère cheese most of which produced heat-labile antibiotic substances able to inhibit growth of and gas formation by *C. butyricum.* An unnamed lactobacillus, isolated by BROSSOWICS *et al.* (1949), produced an antibiotic substance during two to four days of incubation which inhibited *S. dysenteriae, Salmonella paradysenteriae, P. vulgaris, S. typhosa, S. paratyphi, S. shottmuelleri, E. coli, Klebsiella ozenae, Brucella abortus, Pasteurella pseudotuberculosis* and *pestis, Vibro comma, P. aeruginosa, C. diphtheriae, S. aureus,* and *S. pyogenes,* and *faecalis.*

d) Propionibacteria

The propionibacteria, able to produce propionic acid from lactates, contribute to the flavor of Swiss cheese. Eye formation in this cheese is also associated with these organisms. *P. shermanii* has traditionally been employed in Swiss cheese manufacture, although other species in this genus might also be applicable. SARTORY *et al.* (1951) observed that the propionic acid bacteria produced a substance other than the acid which, when at an acid or neutral pH, inhibited the tubercle bacilli. VIERLING (1958 a and b) reported that living cells as well as culture filtrates and cell autolysates of propionibacteria were able to inhibit other microorganisms. The antibiotic was found to be effective in the treatment in mice of typhus and tetanus infections. Certain cancer cells also appeared to be inhibited by the substance. VIERLING (1958 a and b) suggested use of *Propionibacterium freudenreichii, rubrum,* and *technicum.* These propionibacteria were grown anaerobically for up to ten days in a carbohydrate medium at 30° C. This was followed by reduction of the pH to 3.5 with sulfuric acid, vacuum concentration, and extraction with ethyl ether. Evaporation of the solvent yielded the antibiotic which was injectable and could be suspended in water or alcohols.

e) Other bacteria

RITTER (1956 a) reported that *L. helveticus* used to make Swiss cheese was inhibited by certain micrococci found in milk. He postulated that the presence in milk of large numbers of these bacteria could have a disturbing effect on cheese making operations.

Later RITTER (1959) found bacteria of the coli-aerogenes group able to produce substances inhibitory to *L. helveticus* and *lactis* and *S. thermophilus* and *lactis*. Strains of *E. coli* appeared most active against the lactic streptococci while cultures of *A. aerogenes* tended to inhibit the lactobacilli. The inhibitory substance was found to be heat resistant. RITTER (1959) postulated that formation of these inhibitory agents could retard acid formation in cheese milk and thus lead to early gas defects in the finished cheese. He suggested further that retarded acid development without the gas defect in cheese could also be caused by this substance.

BERNHARD (1964) recovered an antibiotic from *Bacillus laterosporus* which he found to be active against *B. cereus* and its variants. This antibiotic, when added to milk containing *B. cereus*, retarded sweet curdling for 1.5 to 4.0 hours. The author concluded that the antibiotic, when combined with sublethal heat treatments, was unsuitable for use in pasteurized milk to inhibit growth of *B. cereus*. FAVILLE and FABIAN (1949) observed that soil forms of aerobic sporeforming bacteria present in vats of fermenting cucumbers produced antibiotic substances able to inhibit *L. plantarum*, the organism chiefly responsible for the fermentation of cucumbers.

An antibiotic able to inhibit *P. vulgaris* was recovered from *Acetobacter acetigenum* by STEEL and WALKER (1958). Bacteria in the genus *Acetobacter* are commonly employed in the manufacture of vinegar since they can oxidize alcohol to acetic acid. Yield of antibiotic was not influenced by the nitrogen source in the medium but was related to the carbon source. The highest yield was obtained when the organism grew for ten days at 30° C. in a medium containing glucose, yeast extract, and peptone. The antibiotic was found to be alkali-labile. Its activity was unaffected by treatment with calcium carbonate, charcoal, and cation- or anion-exchange resins but was decreased by extraction at pH 3.0 with *n*-butanol, chloroform, or methylethyl ketone.

f) Cultured dairy products

Fermented milks have long been a component in the diets of people in various parts of the world. Folklore, practical experience, and some scientific evidence have associated healing as well as nutritive qualities with these products. Yogurt has received more attention in recent years than similar products probably because of its more widespread distribution, palatabality, and the claims occasionally made for it. It is reasonable to assume that if the bacteria used in the manufacture of these products can produce antibacterial substances then the products, too, will contain them. This has been demonstrated repeatedly and will be discussed in some detail below.

1. Yogurt. — This fermented milk product is commonly produced through the joint efforts of *L. bulgaricus* and *S. thermophilus*. SENECA *et al.* (1950) exposed a number of bacteria to yogurt whey for periods of one to 24 hours. After one hour *S. typhi, paratyphi,* and *paradysenteriae, V. comma, B. abortus,* and *B. subtilis* were killed. A two-hour exposure was accompanied by the destruction of *Salmonella pullorum* and *dysenteriae, P. vulgaris,* and *S. aureus.* In four hours *Brucella melitensis* and *P. pyocyaneous* were killed while a five-hour exposure eliminated *E. coli* and *Klebsiella pneumoniae.* After 24 hours *C. diphtheriae,* and *Streptococcus mitis, lactis, faecalis,* and *hemolyticus* were destroyed, while *B. anthracis* and *M. tuberculosis* remained viable. *Entamoeba histolytica,* a pathogenic protozoan, was also inhibited by yogurt whey as were two other protozoa, *Trypanosoma cruzi* and *Leishmania donovani.* These authors found the antibiotic activity of yogurt to be directly related to its pH but concluded that a substance other than lactic acid was responsible for its inhibitory action.

DANON *et al.* (1960), found that *E. coli,* when grown in yogurt, lost its viability in several hours to six days, depending on temperature and acidity. *In vivo* tests with pigs demonstrated that pathogenic and "normal" coliform bacteria were eliminated from the intestinal tract more rapidly when the animals received yogurt than in its absence from their diets. GÜRSEL (1957) claimed yogurt was able to destroy different strains of the human and bovine varieties of *M. tuberculosis* within eight hours although the avian strains of the tubercle bacillus were unaffected. This observation was verified by TACQUET *et al.* (1961) who felt lactic acid was the inhibitory agent in yogurt. Yogurt fed to white mice was found to prevent them from becoming infected by *Erysipelothrix rhusiapathiae* (KATRANDZHIEV 1959). The author suggested that yogurt might be used to control erysipelas in swine.

The optimum temperature for antibiotic production in yogurt was found to be 37° C. by EMANUILOFF and NATSCHEFF (1956). They also tested incubation at 30, 42, 45, and 50° C. No reduction in activity of yogurt serum was encountered when it was heated to 85° C. for one hour or to 120° C. for 150 minutes. Saliva, stomach juices, or pepsin failed to inactivate the serum during a 24-hour exposure at 37° C. while neutralization to pH 7.0, but not to pH 5.5, did.

SCHMIDT and HAUNEMANN (1959) prepared some modified yogurts designated as "Bioghurt" and "special jogurt". They found these preparations able to inhibit laboratory cultures of *Salmonella* and *Escherichia* and that activity was greater at 37° than at 22° or 4° C. The duration of activity in the fermented milks was noted to be inversely proportional to their storage temperatures.

2. Acidophilus milk. — TOMIC-KAROVIC (1962) noted that acidophilus milk had an inhibitory effect on pneumococci and reduced the number of deaths in mice receiving an intraperitoneal injection of pneumococci isolated from positive sputum. Destruction of pathogenic strains of *E. coli* by acidophilus milk was demonstrated by TOMIC-KAROVIC and FANJEK (1962). *In vivo* tests with infants harboring diarrhea-producing strains of antibiotic-resistant *E. coli* indicated the organism was eliminated from the intestinal tract within one-to-five days after acidophilus milk was administered.

3. Other fermented milks. — Kurunga is a fermented milk made with a starter composed of *S. lactis, L. acidophilus,* and undesignated lactose-fermenting yeasts. CHUZKOVA (1958) found it inhibitory to bacilli such as *B. subtilis* and *cereus* var. *mycoides.*

Kumiss is another milk product produced by a combined lactic acid — alcohol fermentation. SKORODUMOVA (1956) found this product able to inhibit *S. marcescens, A. aerogenes, B. subtilis* and *cereus* var. *mycoides, E. coli, Micrococcus candicans,* and two mycobacteria, and postulated that the antibiotic action was the result of symbiosis between the lactic acid bacteria and yeasts. Highest titers were observed when the product was three days old. BUKANOVA (1952) found kumiss and kefir (a similar fermented milk) able to inhibit *E. coli, S. aureus,* and *B. subtilis.*

Swedish ropy milk is a sour milk product made with ropy (i. e., encapsulated, slime producing) strains of *S. lactis.* NILSSON and NILSSON (1955) observed this product had an above average keeping quality and was seldom, if ever, spoiled by molds. Tests with the ropy *S. lactis* culture demonstrated it produced a substance able to inhibit 15 species of *Penicillium, Aspergillus niger,* and one species of *Mucor. Aspergillus flavus,* however, was unaffected. The inhibitory activity appeared to be associated with the slime itself.

g) Cheeses

Production of nisin in or during the manufacture of certain cheeses has been considered earlier in the present paper. Some cheeses are ripened with the aid of molds or bacteria (principally *Brevibacterium linens)* growing on the surface. The presence of antibiotics resulting from the action of ripening agents in such cheeses is a distinct possibility and has, in some instances, been demonstrated.

BINDER (1954) showed that strains of *P. camemberti* (used to ripen Camembert cheese) when grown on laboratory media could effectively inhibit some gram-positive bacteria, *B. linens,* and lactic streptococci. Lactobacilli, micrococci, and *Bacillus* spp. were generally unaffected. The substance was ineffective against gram-negative bacteria and *Geotrichum candidum.* Strains of *P. roqueforti* (used to ripen blue and Roquefort cheeses) tested by BINDER (1954) failed to produce an appreciable quantity of an antimicrobial agent. WILKOWSKE and KRIENKE (1954) assayed 14 mold-ripened cheeses including the blue, Roquefort, Camembert, and Nuworld (made with a white mutant of *P. roqueforti*) varieties for antibiotic substances and were unable to recover any. This was verified more recently by LEDFORD and KOSIKOWSKI (1961).

Inhibition of growth and toxin production by *Clostridium botulinum* and of growth of several penicillia by surface ripened cheese after storage for eight weeks or longer at 2° to 4° C. was observed by GRECZ *et al.* (1959 a). Initially the authors (GRECZ *et al.* 1959 a and c) felt this inhibition was related to presence in cheese of free fatty acids. Further tests (GRECZ *et al.* 1959 b) revealed that *B. linens,* the organism largely responsible for ripening cheeses of the type used in the investigation, produced a substance in cheese extracts and soybean meal broth able to inhibit germination and outgrowth of spores of *C. botulinum* type *A.* The concentration of inhibitor was highest on the seventh day of incubation and decreased slowly thereafter. Heating tests demonstrated the material was stable to an exposure of 120° C. for 25 minutes. Later GRECZ *et al.* (1961 and 1962) indicated the antimicrobial agent a) could be isolated from cheese, b) inhibited *S. aureus* and *B. cereus* in addition to *C. botulinum,* c) could be dialyzed through cellophane casing and adsorbed on Norit *A* at pH 3.0, d) was stable in solutions at pH values of 2.0 to 12.0 when stored at 2° to 4° C., but was destroyed at pH 12.0 by 20 hours of room temperature storage, e) was heat stable especially at lower pH values, f) was soluble in water and common organic solvents, and g) inhibited some gram-positive and -negative bacteria plus some yeasts and molds.

IV. Antibiotics in foods through intentional addition or chance contamination

Use of antibiotics produced by non-food fermentations in the preservation of foods has been suggested by many investigators. Often such proposed uses may lead to the presence of residues in the product, although, in some instances, the antibiotic might be partially or completely inactivated by a heat treatment given the food before it is consumed. Antibiotics have been used extensively in veterinary medicine and as feed supplements. Hence the possibility exists that certain animal products may become contaminated with the chemicals. Use of antibiotics in food preservation and chance contamination of foods will be explored below.

a) Human (mother's) milk

1. Preservation. — Considerable interest has been shown, especially in Germany, in the preservation of mother's milk for subsequent feeding to infants born prematurely or to those who are hypersensitive to cow's milk. According to LINNEWEH (1949 a) any method of preservation must prevent multiplication of coliform bacteria in the milk during refrigerated storage and must also inhibit the lactic acid fermentation. These objectives can be accomplished by heat sterilization but the application of heat reduces the nutritive value of the milk (LINNEWEH 1949 a). Hence, alternative methods of preservation were sought. LINNEWEH (1949 a) found that addition of 2.0 mg.-percent of streptomycin to mother's milk and storing it in previously sterilized glass bottles achieved the desired goals. POETSCHKE (1949) demonstrated that raw mother's milk could be preserved for several months with 2.0 mg.-percent of added streptomycin and refrigerated storage. The milk was found to be free of gram-negative pathogens such as *E. coli, A. aerogenes, P. fluorescens* and *aeruginosa,* and *Proteus* spp. In a later report on clinical experience, LINNEWEH (1949 b) found that the stool flora (*Lactobacillus bifidus* and coliforms) of infants receiving streptomycin-milk remained essentially unchanged. He indicated that the streptomycin would not be detrimental to the infant since it was not absorbed in the intestinal tract. Similar observations were made by POETSCKE (1950).

ROOS and KINDLER (1950) found mother's milk preserved with streptomycin or fortified with 600 mg.-percent of citric acid equally suitable for use. In a later paper ROOS (1951) expressed concern that neither method of preservation would destroy the tubercle bacillus if it occurred in milk.

BINDEWALD (1952) reported that the use of 2.0 mg.-percent of streptomycin was adequate to preserve mother's milk even in warm months. He indicated further that use of streptomycin-treated milk was not associated with intestinal difficulties or hypersensitivity but that use of citric acid-treated milk caused a variety of problems, intestinal and otherwise. Similar observations on citrated milk compared to that preserved with streptomycin were reported by THOMAS (1953).

In later work KAYSER (1952) cautioned against the use of mother's milk preserved by streptomycin since it might contain pathogenic bacteria, and DAMEROW (1956) demonstrated that streptomycin-resistant coliform bacteria may develop in such milk. More recently SCHMIDT *et al.* (1960) demonstrated that the antibiotic activity of streptomycin and dihydrostreptomycin diminished in mother's milk during storage. The authors believed this inactivation resulted from interference by milk proteins and not from chemical decomposition.

OCHLITZ and SCHMIDT (1951) used chlortetracycline to preserve mother's milk and believed it to be as suitable as streptomycin.

2. Contamination. — Antibiotics may be administered to lactating human females for the treatment of an infection. This may lead to contamination of their milk. ROZANSKY and BRZEZINSKY (1949) treated 13 women each with a single intramuscular injection of 200,000 to 600,000 units of penicillin. The antibiotic, at concentrations up to 0.48 unit/ml., appeared in the milk one hour later. Highest milk concentrations appeared two to four hours after injection and the antibiotic persisted in the milk of some subjects for more than nine hours. In contrast to these results, NEUWEILER and KÄSTLI (1953) found chlortetracycline present in the milk of lactating women for seven days after treatment for puerperal mastitis. Highest antibiotic concentrations were observed three to four days after treatment.

More recently TEIXEIRA and SCOTT (1958) administered 500 mg. of novobiocin followed by four doses of 250 mg. at six-hour intervals to lactating women. Concentrations of at least 5.4 mg. antibiotic/ml. of milk were observed 24 hours after therapy was initiated.

b) Vegetables and fruits

Use of antibiotics has been suggested for the preservation of fresh fruits and vegetables and for reducing the heat treatment required to preserve canned products of this type. Fresh fruits and vegetables will be considered first after which canned items will be discussed.

THOMPSON (1959) found that dipping radishes into a solution containing 50 to 100 p.p.m. of oxytetracycline controlled black spotting and sprouting during storage at 70° F. The use of the same antibiotic to preserve cole slaw was examined by FRANCIS (1960) who found that 25 and 50 p.p.m. minimized discoloration and extended the shelf life especially when the product was subjected to higher temperatures for short periods of time.

Solutions of oxytetracycline hydrochloride, streptomycin sulfate, and polymyxin B sulfate (all at 50 p.p.m.) were tested by CARROLL *et al.* (1957) for their ability to control decay of cucumber, chicory, escarole, and lettuce. Oxytetracycline, the best of those tried, retarded spoilage for 48 hours at 30° C. and longer at 5° and 10° C. Streptomycin was second in effectiveness and polymyxin appeared least effective. Similar results on the use of oxytetracycline to extend the shelf-life of fresh leafy vegetables were reported by KERSEY *et al.* (1954). Antibiotic substances recovered from onions, radish, garlic, horse-radish, and mustard, when incorporated into sand or wood shavings used to pack carrots, were found by GRIMM (1959) to enhance their keeping quality when stored at temperatures up to 6° C.

Dihydrostreptomycin and streptomycin sulfate were found (BONDE 1953) effective in the control of rots caused by *Erwinia atroseptica*. Streptomycin sulfate gave complete control of the soft-rot on potatoes treated by immersing them in a solution of ten p.p.m. of the antibiotic for 30 minutes. Oxytetracycline was found to be less bacteriostatic than the streptomycins and chlortetracycline inhibited potato rots caused by *P. fluorescens*. These results prompted the author to suggest that antibiotic treatments might be beneficial for cut seed potatoes just prior to planting and for potatoes being stored prior to their use as food. However, later work by CATES and VAN BLORICOM (1961) indicated that treatment of potatoes with chlortetracycline hastened rotting. MULLER (1958) found bacteria in the genus *Erwinia* susceptible to streptomycin and neomycin but unaffected by seven other antibiotics.

ASSELBERGS *et al.* (1960) treated prepeeled potatoes with low-dosage gamma irradiation and oxytetracycline (dipped for 30 seconds in 100 p.p.m.). Potatoes so treated and packaged in various films remained in satisfactory condition, bacteriologically, for at least six days. However, color changes and development of a "storage" off-flavor would preclude commercial use of this technique.

Darkening of the butts of prepackaged celery is a major problem investigated by HALL (1962). He found chlortetracycline ineffective in the control of this defect. Chlortetracycline has been found effective in the preservation of miso, a fermented soybean product (TOYOSHIMA 1960), and of wet sweetpotato starch (HIGASHI and KANNO 1961), however.

Antifungal antibiotics have been suggested for use in the preservation of certain fruits and fruit products by SALUNKHE *et al.* (1962). COOPER and SALUNKHE (1963) reported that mycostatin and myprozine, when combined with low level irradiation, controlled mold growth on sweet cherries. The efficacy of various antifungal agents in the preservation of berries was evaluated by AYRES and DENISEN (1958). They observed that strawberries and red raspberries dipped in myprozine solutions had lower numbers of microorganisms after storage at 5° C. for nine to ten days than did controls. Trials with other antifungal agents showed that candidin, ascosin, and nystatin at ten µg./ml. were stimulatory to mold and yeast development. At 20 µg./ml. growth of microorganisms was enhanced with candidin or ascosin but not with nystatin; myprozine and rimocidin were inhibitory at both concentrations. An increasing amount of myprozine was associated with fewer organisms on strawberries, raspberries, and cranberries and with an increase in the proportion of sound berries. Addition of 30 p.p.m. of candidicin, mycostatin, or myprozine retarded fermentation and induction of off-flavors in apple juice according to Do and Salunke (1964).

Pimaricin at levels of 1.25 to 20 µg./ml. was found able to retard yeast spoilage of orange juice for eight weeks when stored at 2.5° to 4.0° C., according to Shirk and Clark (1963). Other studies by Shirk *et al.* (1964) showed that ten and 20 p.p.m. of pimaricin also effectively retarded yeast growth in concentrated (42° and 65° Brix) orange juice held at 40° F. The 20 p.p.m. level was most effective when the concentrates were stored at 50° F. or at room temperature. Presence of pimaricin in orange juice can be detected using a cylinder-plate agar assay in combination with *S. cerevisiae* as decribed by Shirk *et al.* (1962).

The combination of an antibiotic with a mild heat treatment has been suggested as a means to can heat-sensitive foods. Wheaton *et al.* (1957), however, reported that subtilin failed to prevent deterioration in tomato juice when it contained spores of *B. coagulans* var. *thermoacidurans* and was underprocessed. Campbell *et al.* (1959) added nisin and/or subtilin to pea puree, cauliflower puree, whole kernel and cream style corn, and tomato juice. Their data suggest that the use of these antibiotics at 14 p.p.m. permitted a reduction in the thermal process necessary for control of spoilage in these foods. Cameron (1951) records data indicating that subtilin at 20 and 40 p.p.m. concentrations failed to prevent spoilage in canned peas, corn, chicken soup, white potatoes, onions, cauliflower, broccoli, celery, and sweet potatoes inoculated with a flat sour or putrefactive anaerobic bacterium before heat processing at 212° F. for ten to 35 minutes. Denny *et al.* (1961) added tylosin to canned corn and mushrooms infected with *B. stearothermophilus* and *C. botulinum* and concluded that the use of sufficient heat to destroy the botulinus organism plus added tylosin served to prevent thermophilic spoilage of the products.

c) Poultry tissue and eggs

1. Contamination from ingested antibiotics. — Antibiotics may be added in low levels (less than 50 g./ton) to feeds to stimulate the growth of young chickens and turkeys. Higher levels may also be administered in the feed or drinking water to control a variety of diseases (Bird 1961). Administration of these low levels generally does not result in the presence of detectable levels of chlortetracycline or oxytetracycline in edible tissues of chickens. Bird (1961) reports further that feeding of tetracyclines in the range of 200 to 1,000 g./ton has led to deposits in the liver while feeding of 1,000 to 2,000 g./ton was associated with deposition in muscle tissue. Other studies summarized by BIRD (1961) revealed that the diet of laying hens had to contain more than 500 g. of chlortetracycline/ton before the antibiotic appeared in eggs. BOYD *et al.* (1960) fed a high level (1,000 g./ton) of chlortetracycline to broilers. Analysis of tissues after one to 11 days of feeding revealed the presence of 0.49 to 0.88, 1.50 to 3.00, 0.68 to 1.30, and 0.59 to 0.75 µg. of antibiotic/g. of serum, liver, breast, and thigh tissues, respectively. Addition of 0.05 percent terephthalic acid to the ration doubled the uptake of antibiotic by tissues. Uptake was increased even further by reducing the calcium content of the diet fortified with antibiotic and terephthalic acid. Liver from birds fed a diet containing 1,000 p.p.m. of chlortetracycline, bacitracin, penicillin, or streptomycin contained an average of 0.938, 0.983, 0.215, 0.0, and 0.0 µg./g. (p.p.m.), respectively of the antibiotics listed. Light and dark meat contained approximately 50 percent of the values listed above for chlortetracycline, oxytetracycline, and bacitracin. Penicillin and streptomycin did not appear in these tissues. Cooking of the meats destroyed all antibiotic activity. BRÜGGEMANN *et al.* (1963) fed C-14-chlortetracycline to laying hens at the rate of 50 mg./kg. of feed. After receiving this feed for 15 days, birds were slaughtered and organs were examined for residues. Calculations based on radioactivity present suggested that the maximum antibiotic content, in µg./g. of wet weight (p.p.m.) was: liver 4.7, spleen 4.7, kidney 5.0, muscle 7.1, heart 2.9, lung 3.6, empty crop 2.3, empty stomach 4.9, empty intestine 4.6, pancreas 4.0, bones 34.5, and blood serum 2.8. Eggs from hens receiving the diet contained 2.15 to 2.77 µg. of antibiotic/g. of egg white, 1.43 to 1.75 µg./g. of egg yolk, and 12.4 to 33.8 µg./g. of egg shell. LOCKHART *et al.* (1962) fed diets containing 50 g. of penicillin or 100 g. of oxytetracycline/ton to laying hens. During a 42-day test period no antibiotic was detectable in the white or yolk portion of eggs produced by the hens. Laying hens were fed diets fortified with 125 to 1,000 p.p.m. of chlortetracycline by FRYE *et al.* (1958). The antibiotic was not recovered from eggs when hens received diets with 125, 250, and 500 p.p.m. of chlortetracycline. When the diet included 750 and 1,000 p.p.m., eggs contained 0.18 to 0.26 and 0.29 to 0.38 µg. of antibiotic/ml., respectively. Concentrations in the yolk were 0.30 to 0.38 and 0.34 to 0.42 µg./ml., respectively, for the two diets while the egg white was free of antibiotic contamination even when the diet contained 1,000 p.p.m. of chlortetracycline. More recently MEREDITH *et al.* (1965) fed a diet containing 1,000 p.p.m. chlortetracycline or oxytetracycline with and without 0.5 percent terephthalic acid to White Leghorn laying hens. When chlortetracycline alone was fed, raw, poached, and scrambled eggs contained 0.02 to 0.08, 0.00 to 0.03, and 0.01 to 0.3 µg. of antibiotic/g., respectively. Addition of terephthalic acid increased these residues to 0.3 to 0.13, 0.02 to 0.08, and 0.03 to 0.06 µg. of antibiotic/g., respectively. Addition of oxytetracycline to the diet yielded eggs which when raw, poached, or scrambled contained 0.25 to 0.32, 0.00 to 0.21, and 0.01 to 0.09 µg. of antibiotic/g., respectively. The presence of terephthalic acid changed these values to 1.0 to 1.1, 0.08 to 0.32, and 0.01 to 0.09 µg./g., respectively. EOFF *et al.* (1962) noted that egg shells became yellow in color when laying hens received feed fortified with 600 to 800 g. of chlortetracycline/ton.

2. Preservation of poultry meat and eggs. — Interest in the use of antibiotics to reduce spoilage of poultry meat and thus improve its shelf-life goes back more than ten years. KOHLER *et al.* (1954) reported that whole eviscerated or cutup chickens dipped in solutions containing three to 30 p.p.m. of chlortetracyline remained fresh and edible for seven to 14 days longer than did untreated controls. They noted that the antibiotic was absorbed by the poultry tissue and was completely destroyed by common cooking procedures even when present 100 times in excess of the amount needed to delay spoilage. ZIEGLER and STADELMAN (1955) confirmed these findings when they observed a 9-to-14 day extension in the shelf-life of chicken fryer halves immersed for ten minutes in ten, 20, or 40 p.p.m. aqueous solutions of chlortetracycline. Although these studies utilized low (36° to 40° F.) temperatures for storage of poultry after treatment, SHANNON and STADELMAN (1957) demonstrated that chlortetracycline enhanced the shelf-life of chicken meat even if stored at temperatures up to 68° F. ESSARY *et al.* (1958), working under commercial conditions, found use of chilled water fortified with chlortetracycline extended shelf-life of fryers by one day when storage was at 34° or at 38° F. Tests by KRAFT *et al.* (1958) on turkeys revealed a single dip into a 20 p.p.m. chlortetracycline solution extended the shelf-life of uneviscerated birds. A second dip after evisceration further extended their shelf-life. MCKEE *et al.* (1959) noted an improvement in the keeping quality of birds dipped in ten to 20 p.p.m. chlortetracycline solutions. They were able to recover up to 6.25 μg. of antibiotic/g. of skin (6.3 p.p.m.) from treated birds and up to 1.63 μg./g. (1.6 p.p.m.) from deeper tissues. BROQUIST *et al.* (1956) and CAREY (1956), also working under commercial conditions, observed that the presence of ten to 15 p.p.m. of chlortetracycline in chill tanks was accompanied by a marked extension of shelf-life. Feeding the antibiotic in concentrations up to 1,000 p.p.m. to poultry before slaughter had no beneficial effect on the shelf-life of the meat afterward, however. Later work of MCVICKER *et al.* (1959) also carried out under commercial conditions indicated that use of chlortetracycline would not always insure an extended shelf-life of fresh poultry.

Equally beneficial results were obtained from the use of either oxytetracycline or chlortetracycline for dipping chicken meat (STADELMAN *et al.* 1957, JACOBSON *et al.* 1959). The former authors note further that holding chickens in ice for 48 hours before cutting up further extended the shelf-life. Later WELLS *et al.* (1957) demonstrated that the shelf-life of poultry treated with chlortetracycline was superior to that when oxytetracycline or tetracycline were used. These authors also noted the emergence of yeasts and bacteria in the genus *Flavobacterium* on treated birds. MCVICKER *et al.* (1958) and SILVESTRINI *et al.* (1959) verified that oxytetracycline was less effective than chlortetracycline in the preservation of poultry. Use of chlorine in the cooling water also was less effective than the chlortetracycline treatment. A somewhat different experimental approach was used by BARNES and SHRIMPTON (1958) in their evaluation of the tetracyclines for preservation of poultry. The three tetracyclines were incorporated at ten p.p.m. in the ice slush used to cool eviscerated poultry which was subsequently frozen, thawed, and stored at 1° C. All antibiotics were found to be equally effective under these conditions. Spoilage organisms developing on treated birds consisted primarily of yeasts and *Achromobacter* spp., while *Pseudomonas* spp. were the main spoilage organisms of untreated poultry. WALKER and TAYLOR (1960) used chlortetracycline at ten p.p.m. to dip chicken meat prior to freezing. After frozen storage for five months, birds were removed to a temperature of 3° C. The antibiotic was still effective and delayed the onset of spoilage. WELLS *et al.* (1958) suggested that the method of packaging poultry may reduce the efficacy of an antibiotic treatment especially if the oxygen tension in the package is reduced. JIMENEZ *et al.* (1961) also studied the effect of packaging and antibiotic on poultry quality. Best results were obtained with whole birds dipped in chlortetracycline and wrapped in Cry-O-Vac or polyethylene films. Tests by DAWSON *et. al.* (1963) indicated that the use of a pasteurization treatment (five to ten minutes at 135° to 140° F.) followed by chilling in slush ice containing ten p.p.m. of chlortetracycline effectively extended the shelf-life of fryers for about one week.

Incorporation of chlortetracycline into slush ice and cooling birds in this ice for two hours was found to be still another successful technique for extending shelf-life (ESSARY *et al.* 1960); extending the icing period to 12 and 24 hours provided no additional benefits.

MEYER *et al.* (1959) incorporated chlortetracycline alone and together with mycostatin into agar and carrageenin which were used to coat the exterior surfaces of poultry meat prior to storage. Shelf-life was extended most when the antibiotic plus the antifungal agent were incorporated into the coating material.

Application of chlortetracycline to chicken fryers by means of dipping, spraying, or slush-ice was investigated by EKLUND *et al.* (1961). They concluded that the slush-ice procedure was preferable and found increases in shelf-life of 4.0, 5.4, 3.7, and 3.2 days when the ice contained 1.0, 5.0, 10.0, and 15.0 p.p.m. of antibiotic, respectively. Gram-negative bacteria predominated on carcasses exposed to the lower concentrations while yeasts and molds were the principal organisms present when ice contained ten and 15 p.p.m. of antibiotic.

The use of chlortetracycline (the "Acronize" process) to prevent spoilage of poultry has also been described in several popular articles (MILLER 1956, WRENSHALL and MCMAHAN 1956). At least two patents have been issued on the process (WINTERBOTTOM *et al.* 1960 a and b). Treatment of poultry with chlortetracycline had no detrimental effects when meat was fed to albino rats (BRÜGGEMANN *et al.* 1961).

The treatment of poultry meat with an antibiotic is not a substitute for adequate sanitation during processing. This was indicated in investigations by GINSBERG *et al.* (1958) who also felt that the antibiotic treatment would not make the carcass of an unhealthy bird suitable for food use. JANKE *et al.* (1958) suggested that processing of poultry to be treated with chlortetracycline should take place under hygienic conditions in order to minimize contamination of the outer skin surfaces. These authors were of the opinion that success with the antibiotic treatment was dependent largely on sanitation practices employed during processing. Under commercial conditions chlortetracycline failed to lengthen shelf-life of poultry with high initial numbers of bacteria (e. g., produced under unsanitary conditions) (DAWSON and STADELMAN 1960). WALKER *et al.* (1960), in Great Britain, also noted a more rapid deterioration of chlortetracycline-treated birds processed under unhygienic conditions than of those handled in a sanitary fashion. In Great Britain dressed poultry is commonly distributed in the uneviscerated form since refrigeration is not used at all stages between the plant and consumer. A "greening" defect is frequently associated with poultry distributed in this manner. The defect results from hydrogen sulfide formed by multiplying microorganisms in the intestinal tract. Hydrogen sulfide diffuses from the intestine into muscle tissue and, in the presence of air, reacts with blood and muscle pigments to form sulfhaemoglobin. SHRIMPTON *et al.* (1958) attempted to minimize this defect by supplementing with antibiotics drinking water consumed by birds during the 24 hours prior to slaughter. The storage life of carcasses from birds receiving the antibiotic as described was extended appreciably by compounds in the following order of effectiveness: oxytetracycline, chlortetracycline, tetracycline, and bacitracin plus neomycin. This technique was also employed by SILVESTRINI *et al.* (1958) who found that the presence of 500 g. of chlortetracycline/ton of drinking water one day prior to slaughter extended the shelf-life of broilers for approximately 11 days.

Processing of poultry with antibiotics may introduce new problems such as the emergence of resistant strains of bacteria or the proliferation on poultry of yeasts unaffected by the chemical. NG *et al.* (1957) studied cultures of bacteria isolated from poultry processed in plants using chlortetracycline regularly. When compared to cultures recovered from unexposed birds, it became apparent that resistance had developed. Actually the number of resistant cultures ranged from 24 to 78 percent of the total isolated in various trials. WALKER and AYRES (1958) observed the development of resistant bacteria on poultry treated with chlortetracycline, oxytetracycline, and tetracycline and stored for ten to 12 days. Later VAUGHN *et al.* (1960) further verified that repeated exposure to increasing concentrations of chlortetracycline resulted in the development (e. g., selection) of antibiotic resistance in cultures of bacteria recovered from spoiled poultry. WELLS *et al.* (1963 a and b) observed that use of chlortetracycline to extend the shelf-life of fresh poultry meat gives rise to bacterial populations exhibiting an increased lipids content. These authors suggest further that cells with more lipids, especially if they were in the cell wall, could experience impaired absorption and thus exhibit resistance to chlortetracycline.

COREÝ and BYRNES (1963) and BYRNES and COREY (1961) recovered oxytetracycline-resistant paracolons, pseudomonads, *Escherichia* spp., and *Aerobacter* spp. from commercially frozen poultry meat which was not treated with the antibiotic. They suspect that these antibiotic-resistant bacteria arose in the intestines of the chickens as a result of ingesting an antibiotic-fortified feed.

The use of chlortetracycline to treat poultry was studied in Canada by THATCHER and LOIT (1961). They reported a) antibiotic-treated poultry contained psychrophilic and mesophilic bacteria in a ratio of 12 : 1, while untreated birds contained these organisms in a ratio of 28 : 1, b) an increase in numbers of yeasts was associated with antibiotic treatment, c) staphylococci were present in relatively low numbers and increased very little during storage of treated or untreated birds, d) enterococci were regularly present but they multiplied little during storage of treated or untreated birds, e) coliforms multiplied substantially during storage, and f) salmonellae were recovered in fewer instances from treated than from untreated poultry.

NJOKU-OBI *et al.* (1957) isolated fungi from chicken carcasses with and without a previous chlortetracycline treatment. Yeasts isolated from treated meat include: *S. cerevisiae, Saccharomyces dairensis, Rhodotorula minuta, G. candidum, Torulopsis holmii, Candida parapsilosis* (a pathogen), *Torulopsis globosa,* and *Candida guilliermondii.* Those recovered from untreated controls were *R. minuta, G. candidum,* and *Torulopsis albida.* Molds isolated from treated meat were in the genera *Cladosporium, Mucor, Penicillium, Rhizopus,* and *Trichoderma,* while those found present on untreated meat include *Alternaria, Aspergillus, Mucor,* and *Rhizopus.*

WELLS and STADELMAN (1958) also studied the development of yeasts on antibiotic-treated poultry meat and recovered some in the genera *Rhodotorula, Torulopsis,* and *Cryptococcus.* A greater proportion of the treated meat had a flora composed principally of yeasts and hence was spoiled through growth of these organisms. WALKER and AYRES (1959) recovered yeasts in the genera *Rhodotorula, Trichosporon, Torulopsis,* and *Candida* from tetracycline-treated poultry meat. The antibiotics failed to stimulate growth of any yeasts recovered. The use of nystatin to control yeast growth on chlortetracycline-treated poultry has been suggested by YACOWITZ *et al.* (1957). They found addition of five and ten p.p.m. of nystatin plus ten p.p.m. of chlortetracycline effectively prevented growth of yeasts and molds and development of fungal odors on stored poultry meat. Nystatin did not penetrate chicken skin or muscle and was rapidly inactivated by boiling. A patent on the use of chlortetracycline and nystatin in dipping solutions into which poultry is immersed for short periods has been issued (PAGANO *et al.* 1960). A combination of chlortetracycline and myprozine also effectively retarded development of bacteria and fungi (AYRES 1959).

SHIRK *et al.* (1957) found that isochlortetracycline was formed when chlortetracycline-treated poultry was heated. Isochlortetracycline was also demonstrated to be the normal degradation product of chlortetracycline in the intestinal tract of the rat. It was microbiologically inactive and had an oral LD_{50} greater than ten g./kg. in mice.

PHILLIPS *et al.* (1960) combined an antibiotic treatment (ten p.p.m. of chlortetracycline in the tissue) with irradiation (100,000 rads) for the preservation of prepackaged fresh chicken meat. Treated meat was held at 41° F. for seven weeks without microbial spoilage. Results obtained on meat independently treated with either irradiation or antibiotic were inferior to those achieved with the combination process. THORNLEY *et al.* (1960) studied the microflora of poultry treated with both chlortetracycline (ten p.p.m. dip) and irradiation (250,000 rads). The initial flora were found to be the same as that associated with the irradiation treatment alone (i. e., ten percent *Achromobacter* spp. and 90 percent yeasts). The extension in shelf-life was roughly equal to that of the sum for the two separate treatments.

The use of tetracycline for the preservation of shell eggs has been suggested by GIOLITTI and GIANELLI (1960). Eggs were covered with films of a) paraffin oil, b) a mixture of oil, Tween 80, and antibiotic (1.0 mg. per ml.), c) an alcoholic solution of the antibiotic mixed with citric acid, or d) oil, alcohol, and antibiotic and held at 18° to 35° C. for eight months. Fifty to 80 percent of antibiotic-treated eggs were sterile while all of those covered with a film of oil only were contaminated. Earlier tests by COTTERILL and HARTMANN (1956) showed no beneficial effect derived from dipping washed eggs in solutions of chlortetracycline, chloramphenicol, neomycin, polymyxin, or streptomycin. MILLER and MORRISON (1958) found that eggs dipped in a 20 p.p.m. solution of chlortetracycline exhibited a higher incidence of spoilage than did controls. A similar observation was made by SCHMIDT and STADELMAN (1957). ELLIOTT and ROMOSER (1957) determined the antibiotic residue in eggs after dipping in solutions of 125 to 1,000 p.p.m. of chlortetracycline. Albumin contained an average of 0.17 to 0.93 μg./g. (0.2 to 0.9 p.p.m.) while the average contents of the membrane and shell ranged from 0.08 to 1.21 μg./g. (0.1 to 1.2 p.pm.) and of whole egg from 0.09 to 0.20 μg./g. (0.1 to 0.2 p.p.m.). These investigators also suggested that chlortetracycline (if it penetrates the shell) is "bound" by the albumin and so loses activity. This would explain why, generally, spoilage of shell eggs was not inhibited by the antibiotic. According to KERSEY *et al.* (1954) oxytetracycline in dried eggs and dried egg products effectively controlled *Salmonella* infections although high concentrations of antibiotic were needed.

d) Pork, beef, lamb, and meat products

The use of antibiotics in the diet of swine has been suggested by different authors (HARRINGTON and TAYLOR 1955, KLINE *et al.* 1955, ROBINSON *et al.* 1956, EZDAKOV 1958, PIVNYAK and OBRAZTSOVA 1958, ARPAI *et al.* 1960). Relatively few studies have been reported on the presence of antibiotic residues in pork from swine which ingested the antibiotic. KELCH (1960) fed chlortetracycline at 32 mg./animal/day to swine for a period approximating six months. The antibiotic was present in all instances in the gall bladder and, in some instances, in the kidney and the liver; spleen and muscle tissue, in all instances, were free of the antibiotic. Canned hams were unaffected by incorporation of the antibiotic into the rations of swine.

Information on the addition of antibiotics to pork for its preservation is relatively limited. PROST (1955) found the onset of putrefaction in ground pork was retarded for only 24 hours through addition of penicillin and dihydrostreptomycin, but an added heat treatment served to eliminate putre factive spoilage. Several patents have been issued on processes designed to preserve pork through the use of antibiotics. In one (WRENSHALL *et al.* 1960) a method is described in which hogs, sheep, and beef cattle are injected intraperitoneally with an antibiotic solution of tetracycline, chlortetracycline, or oxytetracycline prior to slaughter. The antibiotic together with citric acid is administered one-half to 24 hours before slaughter. Tissues recovered from a pig treated in this manner one hour before slaughter contained 4.3, 3.0, 37, 41, 0.94, and 4.1 µg. of antibiotic/g. (p.p.m.) of heart, muscle, kidney, liver, fat, and blood, respectively. Values somewhat higher were reported for the sheep tissue. A similar process for use with beef cattle and swine was described in another patent issued to WILLIAMS (1960). Processes in both patents include an aging procedure in which the meat is held at an elevated temperature, 70° to 118° F. and 50° to 120° F., for one to four days. HALLECK *et al.* (1958) observed that penicillin (0.16 mg./g.) and streptomycin (0.4 mg./g.) added to ground pork retarded bacterial development for two days when the meat was held at 40° to 44° F.

Most work on preservation of meats with antibiotics involves beef, except for poultry and fish. As early as 1952 GOLDBERG (1952) reported that addition of penicillin, streptomycin, and bacitracin (0.5 to 2.0 p.p.m.) to ground beef failed to extend its shelf-life over that observed with untreated controls, but the tetracyclines at the same rate doubled shelf-life. Later GOLDBERG *et al.* (1953) demonstrated that at least one antibiotic, chlortetracycline, disappeared from stored beef tissue within 72 hours. Disappearance of this antibiotic from beef tissue was also demonstrated by WEISER *et al.* (1953 and 1954) who infused animals with a chlortetracycline solution after bleeding but before dressing. Infusion of round steaks with 30 to 50 p.p.m. of oxytetracycline prior to cutting permitted aging at 110° F. for 24 hours without concurrent microbial deterioration (WILSON *et al.* 1960). HENNEBERG and KISSLING (1959) found that chlortetracycline permitted successful aging of beef and mutton at 25° to 32° C. for 48 hours, but it was detectable in the meat for up to 22 days after treatment although decreasing in concentration.

JAY *et al.* (1957 a) examined in greater detail the effect of chlortetracycline on the microflora associated with beef. They allowed beef rounds to spoil after which organisms in the following genera were recovered: *Proteus, Micrococcus, Streptococcus, Escherichia, Bacillus, Alcaligenes, Paracolobactrum, Corynebacterium, Streptomyces, Sarcina, Achromobacter, Flavobacterium, Pseudomonas,* and *Clostridium.* Fewer coliforms, enterococci, and bacteria in the genus *Bacillus* were recovered from rounds infused with the antibiotic. Some antibiotic-resistant micrococci emerged on the treated rounds and inhibition of *P. vulgaris* by five p.p.m. of chlortetracycline was reversed by the addition of manganese ions.

Dipping of whole beef carcasses in solutions of tetracycline-type antibiotics retarded spoilage by one to two days (OSHIMA and UJIIE 1959). The 30-minute dip did not seem to result in development of resistant microorganisms.

The use of oxytetracycline in conjunction with the irradiation treatment of beef was found to prolong its shelf-life (NIVEN and CHESBRO 1956).

It was noted with beef (as has been described with poultry earlier) that fungi, especially yeasts, developed in antibiotic-treated meat.

PHILLIPS *et al.* (1961) found a combination of 7.0 p.p.m. of chlortetracycline and irradiation with 0.1 megarad more effective in preserving beef than either treatment by itself. While use of low irradiation doses (0.18 to 0.72 megarad) permitted retention of activity by tetracyclines used to treat beef, an increase to 4.65 megarads was accompanied by a loss of antibiotic activity (CAIN *et al.* 1958). Maximum complementary action between antibiotic and irradiation was observed when the dose of the latter was 0.1 megarad.

JAY *et al.* (1957 a and b) studied the effect of chlortetracycline on certain strains of *P. vulgaris* isolated from beef and found them to be quite resistant to the antibiotic *in vitro* but quite susceptible in meat. They concluded that the antibiotic disturbed the ionic balance of the meat and thereby retarded growth of the organism. The organic compounds investigated had no effect on reversing the inhibition but certain inorganic materials did, as listed in descending order of effectiveness: tungstate, molybdate, magnesium, calcium, borate, silicate, strontium, and barium. Later JAY *et al.* (1958) indicated that resistant bacteria which, when in raw beef, were susceptible to chlortetracycline, again were resistant when added to roast beef. They believed this resistance was related to the release, through cooking, of nutrient substances in the form of anions and cations which were available to the organism. JAY (1960) was able to increase the resistance of *S. aureus* by serially exposing it to beef fortified with low concentrations of chlortetracycline. Resistance had already increased after three exposures with a further enhancement noted after six exposures. Staphylococci with greater resistance were isolated from beef treated with 0.25 p.p.m. of chlortetracycline than from that fortified with 0.50 p.p.m. of the antibiotic.

SLEETH (1960) injected beef cattle prior to slaughter with oxytetracycline and found that it provided protection against microbial deterioration during high-temperature aging only if the injection occurred within the 30-minute period preceding slaughter. Spraying or washing of carcass surfaces with an oxytetracycline-nystatin mixture effectively controlled surface slime and other surface spoilage during aging. Some oxytetracycline residue was found in treated steaks cooked to the medium-rare stage but not in those which were medium-well done. MATSUI *et al.* (1959) administered oxytetracycline to beef by intraabdominal and intramuscular injections of cattle before slaughter or by dipping of the carcasses afterward. All methods proved effective in retarding the onset of putrefaction in the meat. Feeding to rats of beef preserved with oxytetracycline had no effect on their intestinal flora (SLEETH *et al.* 1960). Detectable levels of antibiotic failed to appear in blood serum, heart, kidney, or muscle tissues.

The shelf-life of veal has also been extended by applying chlortetracycline to the surface of meat (GOLA *et al.* 1959) or by injecting calves intravenously with 10,000 to 100,000 units of penicillin/kg. of body weight 20 minutes before slaughter (HEJLASZ *et al.* 1957).

Besides that of WILLIAMS earlier mentioned (WILLIAMS 1960), patents have been issued on the use of antibiotics in the preservation of beef. These include methods for addition of chlortetracycline, oxytetracycline, or chloramphenicol-plus-ascorbic-acid to ground beef to minimize color deterioration during storage (Savich and Jansen 1957); injection of an animal with chlortetracycline before slaughter or infusion with the antibiotic afterward to preserve the beef and permit rapid aging (Deatherage 1956); simultaneous injection of an antibiotic (any of the tetracyclines) and a proteolytic enzyme to tenderize and preserve the beef (Williams 1963); injection of an animal with a tetracycline antibiotic with or without an added organic acid followed by aging beef at 50° to 120° F. for one-to-four days (Wrenshall *et al.* 1960, Williams 1960); and treating meat surfaces with a mixture of a tetracycline antibiotic and myprozine, applied as a dust or dip (Kohler *et al.* 1962).

Ayres (1959) has suggested the incorporation of chlortetracycline into pliofilm used to wrap meat. This technique served to reduce bacterial growth on the surface of beef and so extended its shelf-life. Discoloration of cooked fresh ground meat was not affected by subtilin but was prevented by the presence of 500 units of streptomycin/g. of meat.

As might be expected, antibiotics have been suggested for use in the preservation of sausage. Deatherage (1957) suggests that antibiotics can prevent the formation of slime which is often induced by temperature fluctuations during the storage of sausage. Levomycetin, when added to liver and meat sausages at concentrations of 100 to 200 µg./g. effectively extended the shelf-life although a bitter flavor was noted when the highest concentration was used (Kazakov and Dyklop 1955). Lorch *et al.* (1961 and 1963) controlled fermentations in sausages by adding 2 to 4 µg. of tetracycline/g. and inoculating with resistant strains of *L. acidophilus* and *Pediococcus cerevisiae;* both quality and shelf-life of the sausage were improved by this process.

e) Fish and other seafoods

The principal problem in the handling of fish, as in many other foods, is microbial deterioration. Early methods of preservation were based on use of high salt concentrations, reduction in pH, or addition of chemicals which may have been questionable from a toxicological viewpoint. Furthermore, these methods invariably damaged the delicate flavor of fresh fish. More recently the broad-spectrum antibiotics, and specifically the tetracyclines, have been found useful for the preservation of fresh fish of various types. Most work has been done with chlortetracycline although the others have also been used. These antibiotics and others used in conjunction with fish will be discussed below.

1. Chlortetracycline. — This antibiotic has been suggested for the preservation of many different fish under a variety of conditions. Some of these are summarized below:

a) Whale. — When injected soon after harpooning, it controlled growth of bacteria for up to 50 hours in muscle near backbone, meat from the throat, deep meat from the abdominal side, and surface meat near blubber but not in various parts of the liver (Duncan *et al.* 1957); injecting antibiotic into the whale after killing it was also successful (Crean *et al.* 1956).

β) Shucked oysters. — Freshly shucked oysters were rinsed in water containing two to 20 p.p.m. of this antibiotic and then stored at 32° to 50° F. Assay of oysters shortly after treatment revealed 0.55, 0.63, 1.3, and 3.0 p.p.m. of antibiotic when rinsed in water containing two, four, ten, and 20 p.p.m., respectively; these concentrations decreased by 50 to 60 percent during 12 days' storage; the antibiotic did not always suppress bacterial growth but prevented formation of the sour odor typical of spoiled oysters (BOYD and TARR 1956). Addition of 20 p.p.m. to water used to wash shucked oysters prior to icing or frozen storage aided in their preservation (NOVAK *et al.* 1958). Dipping in solutions of five and 15 p.p.m. of this antibiotic failed to extend the shelf-life of oysters stored at 1° C.; after 50 days of storage the oysters contained up to 11.0 μg./g. (11 p.p.m.) and this decreased to 0.91 after breading and frying (BENARDE and LITTLEFORD 1957). Higher concentrations of this antibiotic are needed to preserve oysters and other shell fish than for other types of fish (TARR 1959). Frying reduced the antibiotic activity by 84 percent and stewing by 75 percent (KLINE *et al.* 1960). Use of washes and dips fortified with five to 20 p.p.m. extended shelf-life for three to five days and inhibited growth of most indigenous bacteria (NOVAK *et al.* 1960).

γ) Lingcod. — Bacterial growth at 3° C., was retarded by dipping in a solution of 5.0 p.p.m. of this antibiotic or by storing in ice containing a similar concentration (SOUTHCOTT 1958). Growth of bacteria in minced lingcod muscle stored at 1° and 5° C. was inhibited by addition of 2.5 and 5.0 μg./g. (BOYD *et al.* 1956). Immersion of lingcod in fresh sea water containing 2.0 p.p.m. enhanced keeping quality; assay for antibiotic in fish is difficult since other inhibitory substances are normally present (BOYD *et al.* 1953). Immersion of lingcod for 11 days in sea water containing approximately five p.p.m. resulted in deposition of 1.85 p.p.m. in the skin and 2.25 p.p.m. in the muscle of the fish (BAKER *et al.* 1958). Lingcod steaks did not spoil in nine days at 3° C. or 13 days at 0° C. when previously immersed in solutions containing ten or 20 μg. of antibiotic/ml. (TARR *et al.* 1953). Dipping of lingcod fillets in brine solution containing 2.0 p.p.m. extended their shelf-life (FARBER 1954). Heating at 212° F. for ten minutes reduced the antibiotic content of lingcod from 4.8 to 1.8 p.p.m., while heating for 20 minutes eliminated the antibiotic residue (BISSETT and TARR 1952). Formation of trimethylamine in lingcod was suppressed by dipping fillets in solutions of chlortetracycline (BOYD and SOUTHCOTT 1962).

δ) Crab. — An extension in shelf-life was not obtained by dipping for two minutes in solutions containing up to 40 p.p.m. of this antibiotic although numbers of bacteria remained lower in the treated crab meat (BENARDE and LITTLEFORD 1957). Use of tap or distilled water in preparation of antibiotic dipping solutions had no effect on shelf-life of crab meat (BENARDE 1958). Fried crab cakes of different sizes made from crab meat with up to 20 p.p.m. of chlortetracycline contained antibiotic residues up to 2.64 p.p.m. (BENARDE 1957).

ε) Salmon. — Spoilage of salmon was reduced by storage in brine fortified with 5.0 p.p.m. of this antibiotic or by spraying with solutions containing up to 20 p.p.m.; shelf-life was increased by a factor of 1.5 provided the storage temperature approximated 32° F. (STERN *et al.* 1958). Addition of 1.0 p.p.m. to flake ice used to pack coho salmon inhibited development of bacteria on the fish while in transit to canneries, and similar results were obtained with crushed block ice containing an equal amount of antibiotic (GILLESPIE *et al.* 1954 b). Heating of coho salmon containing 5.1 p.p.m. at 212° F. for ten, 15, 20, and 30 minutes reduced the chlortetracycline content to 2.4, 1.5, 0.6, and 0.45 p.p.m., respectively (BISSET and TARR 1952). Addition of 5.0 μg./g. of minced salmon tissue inhibited growth of bacteria when storage was at 1° C. (BOYD *et al.* 1956). Keeping quality of red spring salmon was enhanced by storage in sea water fortified with 2 μg./ml. (TARR *et al.* 1954). About 0.1 μg. of this antibiotic/g. of fish was recovered from salmon preserved in ice fortified with 5.0 p.p.m. (FIRMAN *et al.* 1956). Antibiotic was inactivated during retorting of canned salmon and smoking of salmon reduced but did not eliminate the antibiotic (KLINE *et al.* 1960). Chlortetracycline penetrated no deeper than three to four mm. from the skin in salmon dipped for 30 minutes in a ten p.p.m. antibiotic-sea water solution; the antibiotic was inactivated by canning, boiling, broiling, and frying (TOMIYAMA *et al.* 1958). Studies by TOMIYAMA and YONE (1959 b) revealed that salted salmon which had been soaked in antibiotic-fortified sea water contained less than 0.1 p.p.m. of chlortetracycline after nine months of cold storage.

ζ) Cod. — The amount of antibiotic absorbed from 25 μg./ml. solution by fish depended on their freshness and on length of time in solution; absorption was most rapid during the first five minutes and ranged from $<$ 0.1 to 0.21 μg./g. (DEVYATKINA *et al.* 1961). Ice fortified with 1.0 μg./g. preserved gray cod while in transit to processing plants (BOYD *et al.* 1956). Gray cod stored in ice fortified with one μg./g. contained up to 0.9 p.p.m. of chlortetracycline ten days later while similar fish stored in sea water fortified with two μg./ml. contained up to 3.0 p.p.m. after seven days; fish flesh containing 6.5 μg./g. was freed of chlortetracycline by holding at 99° C. for five minutes but not by comparable holding time at 82 and 60° C., and fish flesh with 16.0 μg./g. was not freed of chlortetracycline by holding at 99° C. for five minutes (STEINER and TARR 1956). Black cod fillets had a longer shelf-life when dipped into a 2.0 p.p.m. solution of chlortetracycline (FARBER 1954). Treating cod with antibiotic ice or washing them with a chlortetracycline solution doubled storage time; maximum antibiotic recovered from raw fish was 3 p.p.m. and this was virtually eliminated in cooking (SEBASTIO and ORLANDI 1958). Fried gray cod previously stored in chlortetracycline-fortified ice (2.5 μg./g.) contained up to 0.17 p.p.m. of antibiotic while boiled and baked fish was virtually free of residues (BOYD *et al.* 1957). The uptake of antibiotic by cod fillets dipped for 0.25 to 1.0 minute in solutions containing ten to 50 p.p.m. ranged from 0.27 to 2.40 p.p.m. (CASTELL and GREENOUGH (1958 a and b).

η) Halibut. — Skin of fish held for 19 days in sea water fortified at 1.5 p.p.m. contained from zero to 0.74 p.p.m., whereas flesh from the same fish contained zero to 0.14 p.p.m. (SOUTHCOTT *et al.* 1958 a and b). Addition of up to 5.0 μg./g. of minced halibut muscle effectively stopped growth of bacteria when storage was at 1° C. (BOYD *et al.* 1956). 0.80 p.p.m. was recovered from halibut preserved in ice fortified with 5.0 p.p.m. of chlortetracycline (Firman *et al.* 1956).

ϑ) Haddock. — Wrapped haddock fillets pretreated by dipping into a ten p.p.m. antibiotic solution had a shelf-life of seven to nine days while untreated fish was spoiled in five days; wrapping and use of cold storage hastened spoilage when compared to holding fish in ice with or without antibiotic fortification (Castell 1958). Use of chlortetracycline in ice or in dips to treat haddock permitted the continued development of indices of decomposition in some fish graded as satisfactory from an organoleptic standpoint (Hillig *et al.* 1962). Fillets cut from haddock dipped in solutions fortified with four to 80 p.p.m. contained a maximum of 0.13 p.p.m. of antibiotic while those from fish stored in antibiotic ice for 13 days contained 0.21 p.p.m.; dipping haddock in solutions containing eight to 80 p.p.m. extended their shelf-life for one-to-13 days (Castell and Greenough 1958 b). This antibiotic was inactivated during retorting of canned haddock (Kline *et al.* 1960).

ι) Sardine. — Fish treated with this antibiotic could be satisfactorily stored at room temperature for 12 to 60 hours or at refrigeration temperature for 72 to 168 hours (Sebastio and Caracciolo 1957). Treatment of sardines for one hour in a brine containing 20 p.p.m. followed by spraying with a salt brine gave a salted product which, after four days, contained 0.17 p.p.m. (Tomiyama and Sugitani 1959). This antibiotic was inactivated during retorting of canned sardines and canned tuna (Kline *et al.* 1960).

κ) Sole. — The shelf-life of sole fillets was increased by dipping for five minutes in a brine solution containing five or 10 p.p.m., and the uptake by fillets ranged from 0.224 to 1.32 p.p.m. (Farber and Lerke 1957); the uptake by fillets was most rapid during the first five minutes of immersion, the residual antibiotic was stable for seven-to-nine days at 5° C. with losses limited to 16.8 percent, but there were insignificant antibiotic residues of 0.0 to 0.41 p.p.m. in steamed, baked, or fried fillets (Lerke and Farber 1957).

λ) Other fish. — Tomiyama *et al.* (1956 a and b) reported that dipping bottom fish in a ten p.p.m. solution of chlortetracycline followed by storage in a 5.0 p.p.m. solution at —1° to 1° C. increased the shelf-life by up to 560 percent. They observed further that application of ten p.p.m. solution to Pacific round herring enhanced the storage-life by 90 percent at 15° to 20° C. and by 40 percent at —1° to 2° C.; this was verified by Tomiyama (1963). Other studies by Tomiyama *et al.* (1957) demonstrated that chlortetracycline penetrated into tissues of pilchard, small mackerel, and bonito but not into several round fish when all were immersed in chilled sea water fortified with 100 p.p.m. of citric acid and ten p.p.m. of antibiotic. Tomiyama *et al.* (1957 b, c, and d) also observed that 60 minutes at 76° or 93° C. was sufficient to destroy chlortetracycline absorbed by bonito fillets. Freshness of dressed mackerel was best maintained when dipped in sea water fortified with 20 p.p.m. of chlortetracycline followed by sealing in polyethylene tubes (Tsuchiya *et al.* 1959 b). Browning of sun-dried mackerel was minimized (Senn *et al.* 1960) when the fish were treated with chlortetracycline before salting. TOMIYAMA and YONE (1959 a) found 0.05 to 0.09 p.p.m. of antibiotic in the muscle and 0.8 to 1.25 p.p.m. in the skin of mackerel held for one week in ice or sea water containing 3.0 to 6.0 p.p.m. Fresh tunny meat was best preserved by dipping for 30 minutes into solutions containing 20 p.p.m. (TAKASUKA *et al.* 1958). Chlortetracycline when used in ice or applied as dip inhibited development of putrefactive microflora in fresh goby and horse mackerel but was ineffective in the preservation of the anchovy (KOVAL'CHUK 1961). HORIE *et al.* (1961) recovered chlortetracycline-resistant micrococci and pseudomonads from treated sea bass and yellowtail. According to FARBER (1954) use of a 2.0 p.p.m. chlortetracycline dip was ineffective in the preservation of shrimp but did prolong the shelf-life of sablefish fillets.

TOMIYAMA *et al.* (1957 a and b) made kamaboko, or fish-meat jelly, from various fish treated with chlortetracycline. They recovered from 0.03 to 0.11 p.p.m. from the kamaboko and also noted that the skin of cutlass fish, butter fish, croaker, lizard fish, sea eel, and harvest fish contained 0.11 p.p.m. 22 days after an initial dip treatment, with flesh averaging 0.08 p.p.m. Some antibiotic was recovered from boiled lizard fish, broiled sea eel, and fried cutlass fish but not from other cooked fish. Later studies (TOMIYAMA *et al.* 1961) demonstrated that quality was improved more when fish without scales (sharp-toothed eel and hairtail) were dipped in chlortetracycline solutions as compared to those with scales (white croaker and flounder).

YONE *et al.* (1959) observed that soaking butter fish and lizard fish in a ten p.p.m. solution of chlortetracycline or spraying them with a ten-to-20 p.p.m. solution kept them fresher than untreated fish. Residues in the muscle did not exceed 0.1 p.p.m. nor did the antibiotic penetrate more than six mm. below the skin. Ice containing 5.0 p.p.m. of chlortetracycline was used by TSUCHIYA *et al.* (1959 a) to preserve saury which remained satisfactory for 17 days after catching while untreated fish were edible for only eight days; a maximum of 0.34 p.p.m. of antibiotic was detected in fish tissue.

YONE and TOMIYAMA (1960) found that treatment of yellowfin tuna with chlortetracycline while on board ship preserved the fish. Residues in fish tissue did not exceed 0.59 p.p.m. and were reduced by 60 to 80 percent through boiling for three minutes. Similar beneficial preservative action was reported by SASAYAMA *et al.* (1960) who treated cutlass fish with chlortetracycline; residues did not exceed 0.22 p.p.m. UNO *et al.* (1959 a and b) found eviscerated Alaska pollack could best be preserved by storing in ice fortified with 5.0 p.p.m. of this antibiotic after previously dipping them for ten minutes in a 20 p.p.m. solution. Use of a 15 p.p.m. dip prolonged the storage-life of shrimp over controls by about four days, according to LERKE and FARBER (1957). Use of 1.0 p.p.m. in ice had no beneficial effect on the preservation of shrimp, but the presence of ten p.p.m. delayed the onset of spoilage for two-to-four days but was of no value in preventing melanosis, according to FIEGER *et al.* (1956). The use of washes and dips fortified with five to 20 p.p.m. extended shelf-life of shrimp by three-to-five days and inhibited growth of most indigenous bacteria (NOVAK *et al.* 1960).

FIRMAN *et al.* (1956) found that chlortetracycline in ice, as a dip solution, or in freezing brine preserved the freshness of many kinds of fish. They recovered up to 5.5, 1.33, 1.55, 0.98, 1.58, 1.60, 0.30, 2.40, and 0.91 p.p.m., respectively, from scrod, butterfish, sea bass, butterfish, porgy, weakfish, croaker, Boston scrod, and New Bedford scrod given chlortetracycline-dip or -ice treatments.

DESILVA and HUGHES (1960 and 1962) found an increase in free amino acids in the flesh of herring treated with chlortetracycline. A similar increase was noted in untreated fish but was not ascribed to microbial action. Instead, the authors believed the increase resulted from autolysis in the muscle independent of the bacterial flora but that the amino acids so formed could be utilized by bacteria and hence, in time, a decrease would occur.

Use of chlortetracycline in the preservation of fish has associated with it a series of problems. One of these is uniformity of its distribution in blocks of ice. The antibiotic normally tends to concentrate in a rather small area near the bottom center of ice blocks. After incorporating a variety of wetting and chelating agents into water and finding them of little use in effecting uniform distribution, GILLESPIE *et al.* (1954 b) tried substances which produced gels or increased viscosity. They found that addition of 0.1 percent carrageenin to the water-antibiotic mixture resulted in a rather uniform distribution of the antibiotic throughout an ice block. Later MOYER *et al.* 1958 showed carboxymethylcellulose and purified sodium and potassium alginates to be nearly equally effective. These same authors found some inactivation of antibiotic when hard water was used to prepare the ice. The inactivation was associated with a precipitate resulting from the presence of calcium and magnesium salts. BOYD *et al.* (1955 b) established that dry carrageenin can be mixed with propylene glycol which further facilitates its distribution and that of chlortetracycline in ice.

YONE *et al.* (1960) found chlortetracycline to be relatively stable in sea water and attributed this stability to the formation of chelate substances with magnesium, calcium, and other metals. They observed that the bacteriostatic effect decreased in the presence of these metals but this decrease could be reversed by the addition of citric acid. Presence of iron and copper (CASTELL and DALE 1963) also reduced effectiveness of the antibiotic. TAKAHASHI (1962) prevented decomposition of chlortetracycline in sea water by adding nine to ten parts of citric or tartaric acid and 20 parts of hexametaphosphate for each part of antibiotic. Stability of chlortetracycline in solution could also be preserved by addition of fish homogenate or gelatin (KAWABATA *et al.* 1960). CASTELL (1960) established that chlorine inactivates chlortetracycline and hence suggested chlorinated water should not be used for preparation of dip solutions. Later CASTELL and DALE (1963) reported the same observation with iodine.

Combining an antibiotic with an irradiation treatment for the preservation of fish has been suggested by LERKE *et al.* (1961) and TARR (1961 and 1962). LERKE *et al.* (1961) found that subjecting fillets of flatfish and rockfish, salmon steaks, precooked tuna loins, cooked crabmeat, and shrimp to beta radiation in the range of 50,000 to 300,000 rads plus use of chlortetracycline slowed up the onset of spoilage sufficient to warrant practical use. TARR (1962) reported that the combined use of chlortetracycline and gamma irradiation (0.25 to 1.0 million rads) from cobalt-60 extended shelf-life of cod and lemon sole even if they were stored at 16° C. A patent has been issued of the combined use of irradiation and tetracycline antibiotics for the preservation of fish and other foods (*Chas. Pfizer and Co.* 1961).

Chlortetracycline has also been suggested for use to preserve ground fish flesh, and fish sausage. TOMIYAMA *et al.* (1962) found it more effective than other tetracyclines, when used at concentrations below 6.0 p.p.m., to preserve ground fish at 0° C. It became more effective as its concentration was raised from five to 20 p.p.m., but not beyond that level.

Patents issued on the use of chlortetracycline for preservation of fish include those of TARR and GILESPIE (1958), UPHAM *et al.* (1956), and NOVAK (1957). The latter utilized dextran as a carrier for the antibiotic.

Not only has the effect of chlortetracycline on shelf-life of fish been investigated but also its effect on bacteria associated with the fish. BOYD *et al.* (1955 a) added it in amounts up to 5.0 p.p.m. to water which also contained live rock bass; their tests suggested that it had no effect in reducing the number of bacteria in the viscera of living fish. BLUHM and TARR (1956) inoculated minced lingcod flesh with *S. aureus* or *C. botulinum* type *E* and added 2.0 p.p.m. of chlortetracycline. When the fish was stored at 50° F., multiplication of the staphylococcus was retarded and numbers of the botulinus organism decreased compared to the initial inoculum provided the antibiotic was present. BOYD *et al.* (1956) and TARR (1960) believed that the use of antibiotics in fish preservation would not create conditions favorable for the growth of food poisoning organisms.

OKITSU (1960 a and b) found that this antibiotic applied to fish inhibited many bacteria but not *P. fluorescens* and *P. morganii*. The presence of magnesium ions reduced the sensitivity of marine bacteria to chlortetracycline (SOUTHCOTT and TARR 1961); the effect was greatest with tetracycline and oxytetracycline, however. CASTELL and GREENOUGH (1957) found it did not inhibit reduction of trimethylamine oxide to trimethylamine or cysteine to hydrogen sulfide by bacterial enzymes; this was also true of oxytetracycline and nisin.

Fish stickwater, also referred to as press water, is a by-product of the fish-processing industry and consists of the fluids expressed from cooked fish or fish waste materials when the latter are converted to meal. After removal of most of the oil by centrifugation, the stickwater can be concentrated to a product called "solubles" which usually contains 50 percent solids and which must be treated in some manner to minimize the effect of suspended protein on its viscosity when condensed; this protein can be coagulated and removed or hydrolyzed by proteolytic enzymes acting at 140° F. IDLER *et al.* (1955) found a loss in solids occurred through bacterial action when the stickwater was held at 140° F. for 48 hours; this loss was prevented by the addition of up to 10 p.p.m. of penicillin, erythromycin, tetracycline, or hygromycin, and each had appreciable inhibitory action at 1.0 p.p.m.

2. Oxytetracycline. — This antibiotic, too, has been examined for its suitability in the preservation of fresh fish. Generally results obtained with it have been less satisfactory than those observed with chlortetracycline although according to SAKAGUCHI (1960), this antibiotic inhibited toxin production by *C. botulinum* in fish dipped in a ten to 30 p.p.m. solution then stored at 4° C. Specific information obtained from tests with different fish is summarized below.

α) Oysters. — Three minute exposure to five and 15 p.p.m. solutions reduced the viable aerobic population but failed to contribute materially to an extension of shelf-life (BENARDE and LITTLEFORD 1957); fried breaded oysters exposed for three minutes to a 15 p.p.m. solution contained 0.36 p.p.m.

β) Crab. — Dipping for two minutes in up to 40 p.p.m. solutions reduced numbers of bacteria but failed to improve the shelf-life (BENARDE and LITTLEFORD 1957). Use of tap or distilled water to preapre antibiotic dipping solutions had no effect on shelf-life of crab meat (BENARDE 1958). Fried crab cakes made from meat containing up to 20 p.p.m. of this antibiotic contained up to 3.19 p.p.m. (BENARDE 1957).

γ) Salmon. — Storage in brines fortified with up to 20 p.p.m. or spraying with solutions containing up to 20 p.p.m. extended shelf-life by a factor of 1.5 provided fish were held at 32° F. after treatment (STERN *et al.* 1958).

δ) Lingcod. — Dipping fillets in a solution of 5.0 p.p.m. inhibited growth of bacteria for eight but not 12 days; oxytetracycline was less effective than chlortetracycline and about equal to tetracycline in effectiveness (SOUTHCOTT *et al.* 1958 a and b). Formation of trimethylamine was suppresed by dipping fillets in solutions of oxytetracycline (BOYD and SOUTHCOTT 1962).

ε) Other fish. — Oxytetracycline as a 20 p.p.m. dip is effective in the preservation of tunny meat (TAKASUKA *et al.* 1958). The shelf-life of black cod fillets was extended by dipping them into a 2.0 p.p.m. solution of the antibiotic (FARBER 1954); oxytetracycline, however, was less effective than chlortetracycline.

3. Tetracycline. — Studies of this antibiotic for preservation of fish have not been so extensive as have those with chlor- and oxytetracycline. SOUTHCOTT *et al.* (1958 a and b) found dipping of lingcod fillets into a 5.0 p.p.m. solution preserved the fish for eight but not for 12 days. These results were similar to those obtained with oxytetracycline but less satisfactory than when chlortetracycline was used. Tunny meat was well preserved by dipping for 30 minutes into a 20 p.p.m. solution of tetracycline (TAKASUKA *et al.* 1958).

4. Penicillin. — This antibiotic has not been tested extensively for preservation of fresh fish. TARR and DEAS (1948) reported use of ices containing 220 units of penicillin/g. of comminuted red spring salmon and lingcod flesh. DUBROVA (1955) found the combination of penicillin and ecmolin to be more effective in the preservation of fish than was penicillin alone. According to TARR *et al.* (1953) penicillin initially inhibited development of bacteria in brill flesh stored at 3° C. but failed to do so when the fish was 11 days old; in other tests TARR *et al.* (1950) had found penicillin reduced the growth of bacteria on sole stored at 37° F. for seven days if the antibiotic was added at rates of ten to 50 p.p.m.

5. Streptomycin. — Only limited tests have been conducted on the use of this antibiotic in the preservation of fish. TARR and DEAS (1948) reported that incorporation of up to 100 units/g. of comminuted red spring salmon or halibut flesh had little effect on their keeping quality. When added to brill flesh at rates up to 50 µg./g. growth of bacteria was retarded up to seven days after which substantial increases in numbers were observed (TARR *et al.* 1953). Other tests by TARR *et al.* (1950) had revealed that streptomycin reduced growth of bacteria on sole held at 37° F. for seven days provided the antibiotic was added at rates of ten to 50 p.p.m.

6. Oleandomycin. — Limited trials with this antibiotic suggested that for preservation of crab meat dipping in a solution of 10 µg./ml. was equivalent to chlor- or oxytetracycline used at the same levels (BENARDE 1958). Use of this antibiotic was associated with appearance of an off-flavor.

7. Polymyxin. — BENARDE (1958) found this antibiotic less effective than the tetracyclines for preservation of fresh crab meat. He also observed an off-odor in the meat which apparently was associated with the presence of polymyxin. This antibiotic added to coho salmon at levels up to 50 µg./g. retarded bacterial growth for seven days after which a substantial increase in the population was observed (TARR *et al.* 1953).

8. Neomycin. — TARR *et al.* (1953) incorporated this antibiotic into coho salmon tissue at levels up to 50 µg./g. Substantial inhibition of bacteria was noted during the first seven days of storage at 30°C., after which appreciable growth occurred. FARBER (1954) found that neomycin, used as dip at 2.0 p.p.m., was ineffective in prolonging the storage period of fish fillets (lingcod, black cod, sablefish, and shrimp).

9. Subtilin. — Addition of up to 50 µg./g. of halibut flesh had no effect on bacterial growth during six days of storage at 0° C. when compared to untreated controls (TARR *et al.* 1953). Similar results were observed with red spring salmon flesh. TARR *et al.* (1950) also found subtilin to be relatively ineffective for inhibiting growth of bacteria on halibut stored ten days at 33° F. or red spring salmon and sole stored seven days at 37° F.

f) Milk and milk products

Antibiotic residues in milk and milk products may arise from several sources. The drugs may be administered to cows orally (directly or through feed) or by injection (intramuscular, intramammary, etc.). Use of the compounds is generally restricted to control of various diseases but, in some instances, may be added to feed to improve its nutritional qualities. All of these uses may lead to the presence of antibiotics in milk. Contamination of milk in these ways has been discussed previously by MARTH and ELLICKSON (1959 a and b), MARTH (1961 a and b), ALBRIGHT *et al.* (1961), STORGARDS (1962), and MYERS (1964), and hence will not be considered in the present review. Methods used to examine milk for the presence of antibiotic residues have also been described previously by MARTH (1961 b), MARTH *et al.* (1963), STORGARDS (1962), and MYERS (1964) and will not be discussed below. Instead, the sections to follow will be concerned with the use of antibiotics to preserve milk, the presence of residues in various milk products, effects of manufacturing processes on antibiotics in milk, and problems (principally inhibition of starter cultures and public health considerations) resulting from milkborne antibiotics.

1. Antibiotics in the preservation of milk. — The use of streptomycin for the preservation of human milk has been discussed earlier in the present paper. Incorporation of streptomycin and other antibiotics into the product has also been suggested for the preservation of market milk. NAGEL (1950) found that addition of streptomycin to milk retarded the development of *E. coli* and other saprophytes present but at levels used failed to destroy pathogenic bacteria. He also indicated that if the antibiotic is to be used it should be added to milk shortly after it is withdrawn from the udder and before appreciable bacterial development occurs. It was observed that bacteria of the typhus, paratyphus, and enteritis types were not destroyed by up to 20 µg. of antibiotic/ml. of milk. A patent has been issued on the preservation of milk with streptomycin (POETSCHKE 1953).

Use of gramicidin for preservation of milk was suggested by LIZGUNOVA (1952). He observed that 400 p.p.m. delayed coagulation for 4.5 days when milk was held at 25° to 28° C. while 0.04 p.p.m. was without effect. Gramicidin, however, failed to destroy typhoid bacteria added to milk but did not induce toxic symptoms in a cat after it ingested 12 to 50,000 µg.

FOLEY and BYRNE (1950) added three units of penicillin/ml. of raw and pasteurized milk. They concluded the antibiotic was of little value in extending the shelf-life of adequately refrigerated, high-quality milk. They did suggest, however, that use of penicillin might be advantageous if proper cooling facilities were not available.

Japanese workers (INOMOTO and HASHIDA 1952, HASHIDA 1953, HASHIDA and ASAI 1953) evaluated use of tetracyclines and chloramphenicol in extending the shelf-life of milk. Putrefactive bacteria in milk were inhibited by chloramphenicol and chlortetracycline while both antibiotics and oxytetracycline inhibited lactic acid bacteria. Shelf-life of pasteurized milk stored at 30° C. was extended from one to 15 days through addition of 0.01 to 0.02 percent of the different antibiotics. Chlortetracycline was believed to be most effective in preserving raw and pasteurized milk. TEJERINA and PORTOLES (1960) reported that milk fortified with 2.0 µg. of chlortetracycline/ml. remained in satisfactory condition for 91 hours when held at 14° C. OLSON *et al.* (1952 and 1953) found that addition of 0.2 µg. of chlortetracycline/ml. of milk had little, if any, beneficial effect on extending its shelf-life. The use of chlortetracycline to preserve raw milk was investigated in detail by DICKINSON and POE (1959), who conclude a) addition of three to ten p.p.m. of antibiotic served to extend somewhat the shelf-life of initially good milk, b) refrigeration and antibiotic treatment complement each other and together serve to extend significantly the shelf-life of raw milk, c) pasteurization of milk, after addition of small amounts of chlortetracycline, resulted in a significant delay in the onset of spoilage, and d) maximum efficiency in preservation was attained when pasteurization was accomplished shortly after the antibiotic was added. EVANS and CURRAN (1952) added from ten to 100 p.p.m. of subtilin to whole or evaporated milk which was then heated to 100° C. for ten or 20 minutes; this combination treatment failed to inhibit spoilage by *C. botulinum* and hence was considered unsafe.

2. Contamination of milk and milk products with antibiotics. — It has been mentioned earlier that milk may contain antibiotics if it is produced by cows being subjected to an antibiotic treatment. The kinds and concentrations of antibiotics in milk as a result of different treatments given the animals have been discussed by MARTH and ELLICKSON (1959 a), MARTH (1961 a), ALBRIGHT *et al.* (1961), and STORGARDS (1962), and will not be reviewed in the present paper.

a) Antibiotics in milk products. — Use of antibiotic-contaminated milk for manufacturing purposes suggests that resulting milk products would also contain the chemical. This has been verified in some instances although results from large-scale surveys of dairy products are limited. One survey conducted by the *Food and Drug Administration* in 1954 failed to detect antibiotic residues in cheese, butter, dried milk, or evaporated milk (WELCH 1957, WELCH *et al.* 1955). HANSEN *et al.* (1950) showed that antibiotics were present in a dried milk made from a milk mixture which contained one percent milk from cows whose udders were recently treated with antibiotics. MARTH and ELLICKSON (1959 a) indicated that analyses by HAGEN and ELLICKSON showed 0.05 to 0.10 unit of penicillin/ml. of certain wheys and of 0.5 to 2.5 units/g. of some dried skimmilks. RUSOFF *et al.* (1957) found that chlortetracycline in milk was concentrated in the skimmilk fraction.

Several antifungal antibiotics have become available and have been suggested for use in minimizing spoilage of cheese. Undoubtedly such uses would lead to residues in the product. Flavofungin, derived from *Streptomyces flavofungini*, is active against a variety of yeasts and molds and is relatively nontoxic. PULAY and BITTERA (1959) employed a 0.5 percent aqueous suspension of this antifungal agent to prevent formation of mold on the surface of Swiss, Pannonia, Port du Salut, and O'vàr cheeses. It prevented mold growth for 12 to 22 days and, although it possessed a bitter flavor, did not impart off-flavors to the cheeses. Resistant strains of microorganisms able to decompose flavofungin were not encountered.

SHAHANI *et al.* (1959) and WILSON *et al.* (1960) added pimaricin at 25 and 50 µg./ml. and mycostatin at ten and 20 units/ml. to water used to wash cottage cheese curd. The shelf-life of the finished cottage cheese was extended three to ten days through use of these antibiotics when storage was at 40° F. Differences were smaller when the cheese was held at 50° and 60° F. Although the antibiotics were active against both molds and yeasts, molds appeared to be most susceptible. CHANDAN *et al.* (1961) found that pimaricin and mycostatin inhibited a bacterial lipase *(Achromobacter lipolyticum)* by 44 to 100 percent, a yeast lipase *(G. candidum)* by 11 to 74 percent, and mold lipases *(Aspergillus niger* and *P. roqueforti)* by 43 to 100 percent.

β) Effect of processing on milk-borne antibiotics. — After milk leaves the farm it is subjected to various manufacturing processes before it reaches the consumer. Most common of these processes are a) various heat treatments as found in pasteurization, evaporation, and drying, b) freezing, and c) concentrating effects of drying, evaporation, and separation.

The effect of heat treatments on the activity of penicillin in milk is summarized in Table I. The work of these authors indicates that penicillin in

Table I. *Effect of various heat treatments on activity of penicillin in milk*

Treatment		Results	Reference
Temperature (°F.)	Time (min.)		
140	30	5 to 30% inactivation	FOSTER and WOODRUFF (1943)
143	30	Not destroyed	KRIENKE (1950c)
145	30	Not destroyed	KATZNELSON and HOOD (1949)
154.4	10	No effect	STORRS and HIETT BROWN (1954)
160	28 hours	Complete inactivation	SHAHANI et al. (1956)
185	10	No effect	STORRS and HIETT BROWN (1954)
190	60	Not destroyed	KRIENKE (1950)
190	7 hours	Complete inactivation	SHAHANI et al. (1956)
190.4	30	No effect	STORRS and HIETT BROWN (1954)
200	230	Complete inactivation	SHAHANI et al. (1956)
212	15	No effect	VIGUE (1954)
212	30	50% inactivation	VIGUE (1954)
212	60	50% inactivation	HUNTER (1949b)
212	60	75% inactivation	VIGUE (1954)
239	10	Complete inactivation	STORRS and HIETT BROWN (1954)
240	15	Not destroyed	KRIENKE (1950)
250	15	Not destroyed	KRIENKE (1950)
250	15	73% inactivation	KATZNELSON and HOOD (1949)
250	25	Complete inactivation	SHAHANI et al. (1956)
250	30	Slight inactivation	SCHIESSLER (1952)

milk is quite stable to heat and that pasteurization treatments commonly applied to milk and milk products are inadequate to destroy the antibiotic if it is present. Some discrepancies exist among the various results obtained at boiling temperatures or above. The data indicate, however, that a portion of the penicillin activity in milk survived boiling for 60 minutes and autoclaving at 15 p.s.i. steam pressure for 15 to 30 minutes.

Antibiotics other than penicillin which may be in milk are also quite heat resistant. It was reported (KATZNELSON and HOOD 1949, KRIENKE 1950 a and b) that heat treatments of 145° F. for 30 minutes, 190° F. for 60 minutes, or 250° F. for 15 minutes failed to inactive chlortetracycline in milk. Pasteurization of milk which contained antibiotics did not inactivate chlortetracycline, chloramphenicol, streptomycin, or oxytetracycline (SHAHANI *et al.* 1954). SHAHANI (1957 b) reported a reduction of 9.3, 14.2, 18.4, 22.5, and 25.4 percent in the activity of chlortetracycline contained in milks heated to 143° F. for ten, 20, 30, 40, and 50 minutes, respectively; the loss in potency was 13.5, 20.1, 27.1, 32.5, and 38.2 percent when milks were heated at 160° F. for the same times. Later SHAHANI (1958) found that pasteurization at 143° F. for 30 minutes resulted in a 30 percent loss of oxytetracycline activity when 0.84 to 1.0 µg. of this antibiotic was present/ml. of milk. An increase in temperature to 160° F. resulted in a 40 percent loss of activity after 30 minutes of heating. Oxytetracycline in milk was found to be completely inactivated by heating to 160° F. for 190 minutes, 175° F. for 92 minutes, and 185° F. for 60 minutes. Cooking of cottage cheese curd which contained 0.1 or 0.2 unit of penicillin/g. at 49°, 52°, or 55° C. for two hours failed to destroy the antibiotic (SMITH 1961).

Low-temperature storage or freezing of milk is not accompanied by a reduction in antibiotic activity when such holding is of relatively short duration. KERSEY *et al.* (1954) showed appreciable stability for the following in milk stored for seven days at 8 to 10° C.: penicillin, streptomycin, oxytetracycline, chlortetracycline, magnamycin, bacitracin, and polymyxin. Penicillin, streptomycin, and a combination of the two were checked by LAURENZA and PIANESE (1956) for their ability to retard growth of bacteria in milk during storage in a refrigerator and at 22° C.; the combination was most effective at both temperatures. Additional comments on effects of low temperature storage are to be found in the next section. HIBBS and BOYD (1957) reported samples of milk which contained penicillin, oxytetracycline, neomycin, streptomycin, or chlortetracycline could be held frozen for at least 12 weeks without appreciable losses of antibiotic activity.

Use of procedures to concentrate milk might also serve to concentrate any antibiotics initially present. Information in this area is quite limited. HANSEN *et al.* (1950) found that separation of milk containing penicillin or streptomycin did not result in a concentration of either antibiotic in the cream or skimmilk fractions. According to RUSOFF *et al.* (1957) chlortetracycline was concentrated in the skimmilk fraction when milk containing the antibiotic was separated. HANSEN *et al.* (1950) reported antibiotic activity present in skimmilk powder made from a mixture containing one percent skimmilk obtained from udders of treated cows. Samples of commercial dried skimmilk, according to MARTH and ELLICKSON (1959 a), were found to contain up to 2.5 units of penicillin/g.

γ) Interaction of antibiotics and milk components. — Antibiotics in milk may affect the enzymes naturally present or may be affected by other milk constituents.

The effect of chlortetracycline and penicillin on the phosphatase enzyme system in milk was investigated by MORR (1959). He found phosphatase was unaffected by chlortetracycline (up to 25 µg./ml.) and only slightly by penicillin (up to 25 units/ml.). Enzyme inhibition produced by penicillin was not enough to alter reliability of the phosphatase test. The degree of enzyme reactivation in milk heated to 210° F. for ten seconds or 145° F. for 30 minutes was not altered by either of these antibiotics. In tests by CHANDAN and SHAHANI (1960) and SHAHANI and CHANDAN (1962) on effect of different antibiotics (up to 50 p.p.m.) on milk lipase, the enzyme was inhibited nine to 42 percent by chlortetracycline, seven to 49 percent by penicillin, nine to 39 percent by streptomycin, and seven to 44 percent by oxytetracycline. These antibiotics were not selective in their inhibitory effects on sites of enzyme action.

The influence of milk components on antibiotic activity was investigated by PRICE *et al.* (1957), MARTIN and HARPER (1959), and HARPER (1960). The former authors showed the antagonistic influence of milk on oxytetracycline, chlortetracycline, and polymyxin could be attributed to the presence of a high concentration of inorganic ions, principally calcium and magnesium. Readily ionizable salts of calcium and magnesium were more inhibitory than poorly ionized salts. Dihydrostreptomycin and neomycin were inhibited by the inorganic ions and by casein. Magnesium salts were found more antagonistic to oxytetracycline and chlortetracycline, and calcium salts were more antagonistic to polymyxin, dihydrostreptomycin, and neomycin.

The loss of antibiotic activity in milk was markedly different for penicillin than for tetracyclines, according to MARTIN and HARPER (1959). The decrease in penicillin activity immediately after its addition to milk or whey was small but considerable inactivation occurred during refrigerator storage. The major loss of oxytetracycline and chlortetracycline activity occurred immediately after addition to either raw or pasteurized skimmilk with only slight losses during cold storage of the milk-antibiotic mixture. The immediate loss of oxytetracycline and chlortetracycline activity was much less in whey than in skimmilk although losses during refrigerator storage were considerable. Dialysis experiments showed no interaction between penicillin and protein systems of either whey or skimmilk. There was interaction between the protein systems of both whey and skimmilk and chlortetracycline with a resultant loss of antibiotic activity in skimmilk only. Radioactive tracer studies showed activity of penicillin was decreased by α-casein and especially so by aged solutions. Calcium ions together with α-casein caused a considerable reduction in activities of oxytetracycline and chlortetracycline. Purified β-lactoglobulin had no significant effect on penicillin or chlortetracycline activity and no chemical interaction was observed between this protein and the antibiotics.

Experiments on the influence of penicillin on hydrogen peroxide added to milk showed the bactericidal efficiency of the peroxide to be reduced by 30 to 70 percent. Interference with the peroxide was proportional to the concentration of penicillin but it was not directly decomposed by the antibiotic (BABAD *et al.* 1959).

δ) *Effects of antibiotics on starter cultures.* — The largest single problem created for the dairy industry by the presence of antibiotics in the milk supply is the slow or complete absence of acid production by bacterial starter cultures employed in the manufacture of a variety of products. The results of studies on concentrations of different antibiotics needed for the partial or complete inhibition of various pure or mixed starter cultures are tabulated in Tables II, III, IV, and V.

Results obtained by different workers on levels of penicillin needed in milk for the partial or complete inhibition of mixed or pure dairy cultures are summarized in Table II. The following conclusions can be drawn from the data: a) *S. lactis* and *S. cremoris* appeared to be more resistant to penicillin than *S. thermophilus,* b) of all the pure cultures listed, *S. thermophilus* seemed to be least resistant to penicillin, c) the lactic streptococci were partially inhibited by 0.025 to 1.0 unit of penicillin/ml. of milk, d) *S. durans* was completely inhibited by the presence of 0.1 unit/ml., e) the lactobacilli were more resistant to penicillin than the streptococci, f) *L. lactis* and *L. helveticus* were less resistant to penicillin than most strains of

Table II. *The concentrations of penicillin in milk needed to inhibit different pure and mixed dairy cultures*

Culture	Level needed for inhibition	
	Partial (units/ml.)	Complete (units/ml.)
Streptococcus cremoris[a]	0.05—0.10	0.10—1.0
S. durans[b]	—	0.10
S. lactis[c]	—	0.05—0.50
S. thermophilus[d]	0.0017—0.17	0.025—0.05
Lactobacillus acidophilus[e]	0.30—0.60	—
L. bulgaricus[f]	0.30—0.60	0.05—0.30
L. casei[g]	0.30—0.60	0.05—2.0
L. helveticus[h]	>0.3	0.10—1.0
L. lactis[i]	—	0.05—0.50
Leuconostoc citrovorum[j]	0.4—0.8	1.6—>1.6
Micrococcus - 8406[k]	—	0.05
Propionibacterium shermanii[l]	0.1	—
S. lactis + *S. cremoris*[m]	0.012—0.096	—
S. lactis + *L. citrovorum*[n]	—	0.1
S. lactis + *L. dextranicum*[o]	—	0.1
S. lactis + *L. bulgaricus*[p]	—	1.0
Butter culture[q]	0.017—0.17	—
Buttermilk starter[r]	0.1	—
Cheese starter[s]	0.05—0.20	0.02—10.0

[a] HUNTER (1949), KÄSTLI (1948), KATZNELSON and HOOD (1949), MEEWES and PASLAWSKI (1951).
[b] SHAHANI and HARPER (1958).
[c] KÄSTLI (1948), MEEWES and PASLAWSKI (1951), SHAHANI and HARPER (1958).
[d] HARGROVE *et al.* (1950), MEEWES and PASLAWSKI (1951), MOELLER-MADSEN and BENDTSEN (1956), SHAHANI and HARPER (1958 b).
[e] WILKOWSKE and KRIENKE (1951 b).
[f] SHAHANI and HARPER (1958 b), WILKOWSKE and KRIENKE (1951 b).
[g] MEEWES and PASLAWSKI (1951), SHAHANI (1957 b), WILKOWSKE and KRIENKE (1951 b).
[h] AUCLAIR (1952), KÄSTLI (1948), MEEWES and PASLAWSKI (1951).
[i] SHAHANI and HARPER (1958).
[j] KATZNELSON and HOOD (1949).
[k] SHAHANI and HARPER (1958).
[l] HARGROVE *et al.* (1950).
[m] BERRIDGE (1956).
[n] HARMON (1951).
[o] HARMON (1951).
[p] SHAHANI and HARPER (1958).
[q] MOELLER-MADSEN and BENDTSEN (1956).
[r] KRIENKE (1950 a).
[s] BAUMGARTNER (1947), DUMAIS (1951), JOHNS (1953), KATZNELSON and HOOD (1949), SCOTT (1950), STOLTZ and HANKINSON (1953 a), TREMBATH (1950), WHITEHEAD and LANE (1956).

L. casei, L. acidophilus, and *L. bulgaricus,* g) the lactobacilli were partially inhibited by 0.03 to 0.6 unit/ml. of milk and completely inhibited by 0.05 to 5.0 units/ml., h) there was considerable variation in penicillin resistance

Table III. *The concentration of chlortetracycline in milk needed to inhibit different pure and mixed dairy cultures*

Culture	Level needed for inhibition	
	Partial (μg./ml.)	Complete (μg./ml.)
Streptococcus durans[a]	—	0.20
S. lactis[a]	—	0.05
S. thermophilus[b]	0.001—0.01	0.03
Lactobacillus bulgaricus[a]	—	0.3—5.0
L. casei[a]	—	0.05
L. lactis[a]	—	0.3—3.0
S. lactis + L. citrovorum[c]	—	10.0
S. lactis + L. dextranicum[c]	—	10.0
L. lactis + L. bulgaricus[a]	—	10.0
Butter culture[d]	0.01—0.1	—
Cheese starter[e]	0.02—0.25	0.25—1.0

[a] SHAHANI and HARPER (1958).
[b] MOELLER-MADSEN and BENDTSEN (1956), SHAHANI and HARPER (1958).
[c] HARMON (1951).
[d] MOELLER-MADSEN and BENDTSEN (1956).
[e] BRADFIELD (1949), BRADFIELD (1950), JOHNS (1953), JOHNSTONE (1950), KRIENKE (1950 a and b), OVERBY (1952), STOLTZ and HANKINSON (1953 a).

Table IV. *The concentration of streptomycin in milk needed to inhibit different pure and mixed dairy cultures*

Culture	Level needed for inhibition	
	Partial (μg./ml.)	Complete (μg./ml.)
Streptococcus thermophilus[a]	0.05—5.0	—
Propionibacterium shermanii[b]	5.0	—
S. lactis + L. citrovorum[c]	—	10.0
S. lactis + L. dextranicum[c]	—	10.0
Butter culture[d]	0.1—0.2	—
Cheese starter[e]	0.04	—

[a] HARGROVE *et al.* (1950), MOELLER-MADSEN and BENDTSEN (1956).
[b] HARGROVE *et al.* (1950).
[c] HARMON (1951).
[d] MOELLER-MADSEN and BENDTSEN (1956).
[e] OVERBY (1952).

between strains of the same species of *Lactobacillus,* i) *P. shermanii* was less resistant to penicillin than the lactobacilli and *L. citrovorum* was approximately equal to the lactobacilli in resistance, and j) resistance of mixed cultures ranged between that of the streptococci and the lactobacilli.

The effect of low penicillin concentrations in milk on acid production by different cheese cultures was investigated by RICHARDS (1958). He found that 0.1 unit/ml. inhibited *S. durans* and *S. thermophilus,* 0.25 unit/ml. inhibited *S. diacetilactis* and *S. cremoris*, and 0.5 unit/ml. retarded mixtures of *S. cremoris* plus *Leuconostoc* spp. and *S. diacetilactis* plus *S. cremoris*.

Data obtained by JACQUET (1959) through experiments on making cheese from penicillin-contaminated milk (0.10, 0.25, and 0.50 unit/ml.) suggested that lactic acid bacteria, during a ten-year period, became less

Table V. *The concentration of different antibiotics in milk needed to inhibit pure and mixed dairy cultures*

Antibiotic	Culture	Level of antibiotic needed for partial inhibition (µg./ml.)
Bacitracin[a]	*Streptococcus thermophilus*	0.05—0.10
	Butter culture	0.10—0.20
Chloramphenicol[b]	*S. thermophilus*	0.05—0.10
	Butter culture	0.10—0.20
	Cheese starter	0.02
Neomycin[a]	*S. thermophilus*	0.10—1.0
	Butter culture	0.10—1.0
Oxytetracycline[c]	Cheese starter	0.01
Tetracycline[a]	*S. thermophilus*	0.001—0.01
	Butter culture	0.01—0.10
Tyrothricin[d]	Cheese starter	1000.0

[a] MOELLER-MADSEN and BENDTSEN (1956).
[b] MOELLER-MADSEN and BENDTSEN (1956), OVERBY (1952).
[c] OVERBY (1952).
[d] STOLTZ and HANKINSON (1953 a).

sensitive to penicillin. Secretion of penicillinase or formation of resistant mutants were suggested explanations. JOHNS (1956) demonstrated that production and accumulation of acid by lactic streptococci made them more susceptible to the action of penicillin. Neutralization of acid after it reached 0.35 percent permitted more acid development by *S. cremoris* even though milk contained 0.15 unit of antibiotic/ml. MEANWELL (1962) observed that lactic starters were less affected by penicillin when incubation was at 30° C. instead of at 22° C.

VERLINSKI (1959) reported a commercial starter was made resistant to penicillin over a period of four-to-five weeks by addition, at each daily transfer, of penicillin in increasing quantities from 0.05 to 0.1 unit/ml. The starter, when fully adapted to the antibiotic and incubated at 26° C., retained normal flavor and developed acid in milk with up to 1.0 unit of penicillin/ml. Attempts were made (RICHARDS and KENNEDY 1960) to produce penicillin-resistant cultures. A selection procedure for isolation of naturally resistant mutants and a training program for induction of resistance were employed. Cultures were obtained which grew in the presence of 0.5 unit of penicillin/ml. of milk although acid production by these strains was reduced by 50 percent when compared to sensitive cultures.

Changes in morphology of lactic streptococci are associated with penicillin inhibition. Involution forms of *S. lactis* and *S. cremoris* have appeared in the presence of the antibiotic (HUNTER 1949 b). Antibiotic levels which caused nearly complete inhibition of the cultures produced marked increases in the length of the cells. The cocci emerged as elongated rods, apparently because of difficulty in cell division. Similar morphological changes were also observed by GALESLOOT (1956 a) when *S. lactis* and *S. cremoris* were grown in milk which contained 0.30 unit of penicillin/ml. He also noted that streptococci in yogurt appeared swollen and existed in longer chains when 0.1 unit/ml. of penicillin was present. BAUGHMAN and NELSON (1958) found that cells of *S. thermophilus* tended to form long chains in concentrations of penicillin as low as 0.03 unit/ml. The gram-positive staining reaction of *S. lactis* and *S. thermophilus* was reversed by exposure to various low concentrations of penicillin. Several tests based on this change in morphology of bacteria have been suggested for detection of penicillin in milk (LISKA 1959 a and b, LISKA 1960, WHITEHEAD and COX 1956).

Data by various workers on levels of chlortetracycline needed in milk for the partial or complete inhibition of mixed or pure dairy cultures are summarized in Table III. The results indicate a) streptococci, in general, were less resistant to chlortetracycline than lactobacilli, b) variations in degree of resistance existed between different strains of the *Lactobacillus* species, and c) mixed starter cultures were somewhat more resistant to chlortetracycline than the streptococci and somewhat less than some of the lactobacilli.

Table IV presents a summary of data on concentrations of streptomycin needed in milk for the partial or complete inhibition of mixed or pure starter cultures. The data indicate: a) considerable variation in resistance of cultures and b) mixed starter cultures studied appeared to be more susceptible to streptomycin than were pure cultures.

Limited data on the inhibitory effects of various other antibiotics on cultures is summarized in Table V. From these results it is evident that a) *S. thermophilus* was partially inhibited by the presence in milk of 0.05 to 0.10 μg. of bacitracin or chloramphenicol/ml., by 0.1 to 1.0 μg. of neomycin/ml., and by 0.001 to 0.01 μg. of tetracycline/ml., b) butter cultures were partially inhibited by the presence in milk of 0.10 to 0.20 μg. of bacitracin or chloramphenicol/ml., 0.10 to 1.0 μg. of neomycin/ml., or 0.01 to 0.10 μg. of tetracycline/ml., and c) cheese starters were partially inhibited by 0.02 μg. of chloramphenicol/ml., 0.01 μg. of oxytetracycline/ml., or 1,000 μg. of tyrothricin/ml.

MIKOLAJCIK *et al.* (1960) found a strain of *S. lactis* which was resistant to oxytetracycline, produced low levels of acid, and possessed the ability to "bind" the antibiotic. It was possible to obtain normal acid development by sensitive cultures in milks with one or ten p.p.m. of oxytetracycline if sufficient resistant cells were added first, given time to "bind" the antibiotic, and thus to neutralize its effects. KENNEDY (1960) attempted to improve acid production of the oxytetracycline-resistant culture through addition of various nutrients to milk. Peptide-rich extracts, when added to milk at 0.1 to 0.5 percent, permitted normal acid production by resistant cultures. Effective additives were yeast extract, liver fraction *L*, neopeptone, tryptone, milk-protein hydrolysate, casein hydrolysate, pancreas extract, tryptose, and peptonized milk. Sensitive cultures, however, did not produce acid in antibiotic-contaminated milk when the extracts were added.

Differences between cells sensitive and resistant to oxytetracycline were investigated by MIKOLAJCIK (1960). He found a) sensitive cells able to tolerate higher concentrations of sodium chloride and alkaline substances than resistant cells, b) sensitive cells had a more diffuse cell wall and more prominent electron dense cytoplasm than resistant cells, c) sensitive cells were less exacting in requirements for anaerobiosis, and d) the medium (containing oxytetracycline) in which sensitive cells grew showed an accumulation of glutamic acid, leucine, acetone, and alanine while aspartic and oxaloacetic acids accumulated after growth of resistant cells. MIKOLAJCIK (1960) felt that the bacterial cell wall was the probable site of oxytetracycline action. This observation lead to development of the method for antibiotic "binding" previously described.

SHAHANI (1957 a and 1959) and SHAHANI and HARPER (1956 and 1958 a) studied differences in phosphorus compounds present in oxytetracycline-sensitive and -resistant cells of *S. lactis*. More cellular material was consistently produced by the sensitive strain. Young cells of sensitive strains generally contained higher levels of protein and nucleic acid phosphorus than did similar resistant cells. The nucleic acid content of both types of 24-hour old cells was nearly equal. Resistant and sensitive cultures showed marked differences in their phosphate metabolism. The acid production of the resistant culture was greatly accelerated by the addition of glucose-1-phosphate and fructose-1,6-diphosphate. Studies with P^{32} revealed a much slower uptake of phosphorus by the cells of the resistant culture than by the sensitive cells. Chromatographic analysis indicated that upon development of resistance the organisms were inhibited in their synthesis of sugar phosphate esters. The inability of the oxytetracycline-resistant *S. lactis* to produce acid rapidly appeared related to the failure of the organism to assimilate phosphorus rapidly and to its slow sugar-phosphate synthesis.

Galactokinase and galactowaldenase were detected in *S. lactis* by SHAHANI (1960). He observed further that penicillin and streptomycin had no effect on galactokinase but the enzyme was completely inhibited by chlortetracycline and oxytetracycline. Penicillin was found to inhibit 33 to 50 percent of the galactowaldenase activity. In later studies VAKIL and SHAHANI (1962 b) found that *S. lactis* was able to utilize lactobionate and gluconate and produce varying amounts of lactic, acetic, and formic acids, carbon dioxide, ethanol, and glycerol. Chlortetracycline and oxytetracycline completely inhibited metabolism of the organism while penicillin and streptomycin did so for only 18 to 24 hours. These antibiotics partially inhibited (0.6 to 6.5 percent) the enzyme able to hydrolyze lactobionate. SHAHANI and VAKIL (1962 a) observed that the presence of low concentrations of penicillin and streptomycin altered the formation of metabolites of *S. lactis*. Penicillin inhibited formic acid production in galactose broth, but had no effect on formic acid development in lactose broth and stimulated its formation in glucose broth. The antibiotic also retarded production of ethanol in galactose and lactose fermentations but stimulated it when glucose served as a substrate. The reverse of this condition existed with regard to glycerol production. Streptomycin was less inhibitory when monosaccharides rather than lactose served as the substrate. SHAHANI and VAKIL (1962 b) and VAKIL and SHAHANI (1962 a) demonstrated that lactase, lactose dehydrogenase, Zwischenferment, and 6-phosphate gluconic dehydrogenase were inhibited to varying degrees by penicillin, streptomycin, chlortetracycline, and oxytetracycline. These enzymes were all recovered from *S. lactis*. Hexokinase was unaffected and aldolase was sensitive to all antibiotics except penicillin.

RICHARDS (1961) examined the influence of antibiotics on susceptibility of lactic starters to bacteriophage infections. Sub-inhibitory concentrations of penicillin were shown to decrease the time required for initiation of lysis by the bacteriophage acting on *S. lactis*. Oxytetracycline caused a delay in the onset of lysis while streptomycin had no effect. The magnitude of the effects (penicillin and oxytetracycline) increased as the antibiotic concentration was increased. Acquisition of penicillin resistance by lactic streptococci altered their resistance to infection by bacteriophages. Non-infective phages became infective. Later MIKOLAJCIK and GOULD (1962) revealed that adsorption of bacteriophage by *S. lactis* was somewhat reduced by penicillin but the burst size of the susceptible bacteria was doubled: 84 particles compared to 42. This serves to explain further the observations of RICHARDS (1961).

According to MARTH (1960 a), GELSEY and HAGEN examined the effect of seven antibiotics on three different strains of *P. shermanii* and on four strains of propionibacteria isolated from Samsoe cheese. All strains were completely inhibited by the presence/ml. of 0.5 to 1.0 μg. of oxytetracycline, 0.5 to 1.0 μg. of tetracycline, 0.1 to 5.0 μg. of chlortetracycline, 1.0 to 10.0 μg. of erythromycin, 0.1 to 1.0 μg. of chloramphenicol, ten to 30 μg. of streptomycin, and 0.6 unit of penicillin.

ε) Public health problems. — The public health problems which may be associated with the presence of antibiotic residues in milk and milk products include a) exposure of the consumer to antibiotic-resistant udder pathogens which may be present in unpasteurized milk products, b) development or aggravation of antibiotic hypersensitivity in consumers, c) alteration of the intestinal flora of consumers to thereby bring about a reduction in vitamin synthesis, and d) alteration of the intestinal flora of consumers to permit the establishment and growth of antibiotic-resistant strains of pathogenic microorganisms.

Cheese made from milk obtained from cows previously treated for mastitis by means of penicillin and streptomycin has been incriminated as the source of antibiotic-resistant micrococci and streptococci which caused infections in consumers (ANONYMOUS 1955). JEZESKI *et al.* (1961) found no higher populations of staphylococci in cheese made from milk fortified with sufficient penicillin to inhibit the starter culture than were present in cheese made from normal milk. In both instances milk was inoculated with staphylococci.

The number of coliform bacteria in the intestinal tract of persons was markedly reduced by the ingestion of 500 ml. of milk which contained either 0.1 mg. of streptomycin/ml. or 0.25 μg. of chlortetracycline/ml. (STOLTZ and HANKINSON 1954). The authors have postulated that the antibiotics induced a reduction in numbers of intestinal organisms which would result in the reduction of vitamin synthesis in the intestinal tract.

Broad spectrum antibiotics (the tetracyclines), neomycin, streptomycin, bacitracin, and polymyxin are poor sensitizers and hence their presence in milk is of minor concern to hypersensitive or potentially hypersensitive consumers (WELCH 1957, ANONYMOUS 1959 a). The presence of penicillin in milk, however, is of concern in regard to the hypersensitivity of the consumer (BRYAN 1951, WELCH 1957).

The majority of a representative group of allergists whose opinions were sought regarding the status of the problem of allergic reactions from penicillin-contaminated milk agreed that a) such reactions have occurred and probably have gone undetected in many instances, b) reactions occurred after ingestion of milk although some felt the small amounts of penicillin in milk were harmless to penicillin-sensitive people except in rare, extremely sensitive individuals, c) an oral dose two-to-three times as great as an intravenous one was necessary to produce reactions in sensitive individuals, and d) reactions most commonly encountered were chronic and recurrent urticaria (ANONYMOUS 1959 b). The Council on Drugs of the *American Medical Association* has confirmed the occurrence of allergic reactions in sensitive persons who consumed milk containing small amounts of penicillin. ERSKINE (1958), VICKERS *et al.* (1958), and BORRIE and BARRETT (1961) noted allergic dermatitis in patients who had consumed milk contaminated with penicillin. ZIMMERMAN (1959) reported case studies of four patients who suffered urticaria after ingestion of dairy products. All patients had previously allergic reactions to penicillin. Reactions were cleared up rapidly by injection of penicillinase, an enzyme able to catalyze the hydrolysis of penicillin to penicilloic acid (ROYCE *et al.* 1952), and subsequent prophylactic injections of the enzyme enabled patients to eat dairy products without allergic reactions. Penicillinase was used by FRIEDLANDER (1959) to treat 50 other patients with penicillin reactions. Rapid improvement after intramuscular administration was observed in some cases although others failed to respond. Local pain, swelling, and systemic febrile reactions were commonly encountered after intramuscular injections of the enzyme. The antigenicity of penicillinase is such that sensitization may result from its repeated administration. It has been suggested that cheeses ripened with molds of the genus *Penicillium* might contain penicillin and thus give rise to allergic reactions. Recent tests by LEDFORD and KOSIKOWSKI (1961) failed to detect this antibiotic in Roquefort, Blue, or Bleu cheeses, however.

The presence of antibiotics in milk may also affect results obtained from quality tests applied to the milk. WILKOWSKE and KRIENKE (1951 a) found storage for 72 hours at 10° C. of raw milk fortified with penicillin was not accompanied by an appreciable increase in numbers of bacteria, while STOLZ and HANKINSON (1953 b) suggested an actual decrease in numbers of organisms might occur in such milk. This could lead to erroneous results when conventional quality control tests are employed.

The length of time needed for the reduction of the dye in the methylene blue test for milk quality was increased when concentrations of 0.1 to 0.5 unit of penicillin/ml. of milk were present (HUNTER 1949 a). The reduction time was also prolonged in storage milks which contained antibiotics. JOHNS and DESMARIS (1953) found that concentrations of 0.05 and 0.5 unit of penicillin/ml. of milk caused an appreciable delay in the reduction time of methylene blue when added to such milk. The retarding influence was found greatest when the dye reduction time was longest and hence the resazurin test (three-hour completion time) was less affected than the methylene blue test. The reliability of the phosphatase test used to detect under-pasteurized milk or milk products was not impaired by the presence of antibiotics in milk (CHURCHILL *et al.* 1951).

g) Fermentations

The addition of antibiotics to certain fermentation mashes has been suggested as a means to prevent growth of undesirable contaminants. Use of antibiotics in this manner could lead to their presence in products resulting from the fermentations. Much of the experimental work has centered around the production of beer and wine although some other applications have been indicated.

According to STRANDSKOV *et al.* (1953) and STRANDSKOV and BOCKELMANN (1953) bacteria able to grow in wort during fermentation and in finished beer include gram-positive lactic acid organisms *(Pedicoccus cerevisiae* and *Lactobacillus pastorianus)* and a gram-negative rod-shaped microbe *(Flavobacterium proteus).* STRANDSKOV and BOCKELMANN (1953) showed that the gram-negative bacterium was inhibited during fermentation by the presence of 0.005 µg. of polymyxin/ml. but not by 0.001 µg. A secondary benefit was observed in that the polymyxin appeared to stimulate the yeast fermentation. Concentrations of streptomycin, bacitracin, penicillin, and subtilin in excess of 125 µg./ml. were required to retard the organism while the tetracyclines did so when present at 50 µg./ml. Penicillin was found to be most effective in the control of gram-positive contaminants. Polymyxin, according to HAAS (1955), is suitable for inhibition but not destruction of *P. cerevisiae.* He found chloramphenicol most suitable for inhibition of *L. pastorianus.* Adsorption of polymyxin by common filtering agents was noted. KERSEY *et al.* (1954) found polymyxin to be quite stable in finished bottled beer. Beer with an initial concentration of 3.1 units/ml. did not suffer from a loss of antibiotic activity during a 16-week storage period. The combined use of polymyxin (6.25 µg./ml.) and penicillin (4.0 µg./ml.) has been found beneficial in the preservation of beer (KATO *et al.* 1957). Several patents have been granted on the use of polymyxin together with other antibiotics to minimize growth of contaminating bacteria during fermentation and in finished beer (BOCKELMANN and STRANDSKOV 1957, STRANDSKOV and BOCKELMANN 1960).

Information on use of antibiotics in wine fermentations is quite limited. PEYNAUD and LAFOURCADE (1953) noted that antimycin *A* and actidione prevented further development of yeast in raisin- and sugar-fortified wine, respectively. RIBERAU-GAYON and PEYNAUD (1952) found penicillin, the tetracyclines, and subtilin ineffective in retarding development of yeasts. DAL-CIN (1948) reported that addition of 0.015 to 0.025 percent biamicina (an Italian antibiotic) to must aided in control of the wine fermentation.

Addition of antibiotics is effective in the control of contaminants during the sake fermentation. SATO (1958) found lactobacilli were inhibited by the presence of penicillin, the tetracyclines, or chloromycetin at 20 μg./ml., while *Acetobacter* spp. were retarded by less than 20 μg./ml. of streptomycin or tetracyclines. Similar results on the control of acid-forming bacteria through the use of penicillin in the sake fermentation were reported by AKIYAMA *et al.* (1958).

Addition of antibiotics to grain mash fermentations, as used in the production of distilled alcoholic beverages, at the levels indicated proved effective in the control of bacterial contaminants: penicillin 0.75 to 2.0 units/ml., chlortetracycline 2.0 to 10.0 μg./ml., bacitracin ten to 20 μg./ml., oxytetracycline 20 to 40 μg./ml., and tyrothricin 300 to 500 μg./ml. (DAY *et al.* 1954). Streptomycin was without appreciable effect at a concentration of 1000 μg./ml. Marginal treatment of grain mashes with an antibiotic was found to be dangerous since it tended to enhance production of acrolein by certain strains of lactobacilli. Presence of antibiotics in the mash also prevented destruction of α-amylase and a concurrent reduction in alcohol yield and inhibited acid production.

The use of tyrothricin to control contaminants in fermentations employed to produce yeast cells was suggested by GRAY and KAZIN (1946). They observed that up to 500 p.p.m. of antibiotic had no effect on yeast cell development but did effectively control growth of gram-positive organisms. When yeast grown in the presence of tyrothricin was used to make beer, a normal product resulted. ROZMANOVA (1958) isolated *Bacterium mesentericus,* micrococci, lactic acid bacteria, and myxobacteria from mash used as a substrate for yeast production. Chlortetracycline and chloramphenicol were able to inhibit the contaminants while penicillin and streptomycin proved less effective. Tests by BARTA (1960) on baker's yeast and on the mash revealed the presence of *Acetobacter* spp., *Lactobacillus* spp., and *Clostridium putrefaciens.* The presence in the mash of ten μg./ml. of chlortetracycline completely inhibited development of these organisms. Use of this antibiotic did not affect unfavorably the enzymic activity of the yeast or its ability to ferment sucrose, glucose, or maltose.

h) Other aspects

The effect of food-borne antibiotics on certain food-spoilage and -poisoning organisms has been investigated by some researchers. As might be expected, most interest has centered around spore-forming organisms since they are normally most difficult to destroy in food processing operations. Some experimental work has been done with nonsporeformers such as staphylococci, salmonellae, and coliforms. Both aspects will be considered below.

1. Non-sporeforming bacteria. — The influence of several antibiotics on growth of *S. aureus* (food-poisoning strains) suspended in a custard filling was evaluated by GODKIN and CATHCART (1952). Proliferation of the organism, at 98.6° F., was retarded for 48 to 72 hours by the presence of 100 p.p.m. of subtilin. Spoilage by the natural heat-resistant flora was also retarded for 48 to 72 hours under the same conditions. A greater preserving effect against both pathogenic and non-pathogenic organisms was obtained when 1.0 p.p.m. of oxytetracycline was combined with 100 p.p.m. of subtilin. Small amounts (0.6 to 1.0 p.p.m.) of chlortetracycline and oxytetracycline effectively retarded the growth of *S. aureus* in the custard filling but exhibited no control of the heat-resistant, non-pathogenic flora responsible for spoilage. Penicillin was less effective against *S. aureus* than either chlortetracycline or oxytetracycline and showed very little sporicidal activity. Bacitracin, chloramphenicol, and streptomycin were without benefit at up to 60, 5.0, and 5.0 p.p.m., respectively.

Greenberg and Silliker (1962 c) noted that development of staphylococci in ice cream mix was prevented by the presence of 20 p.p.m. of tylosin. Further, the growth of this organism in processed Cheddar cheese spread, ham, and sausage was prevented by 2.5, 3.0, and 5.0 p.p.m. of tylosin, respectively. These authors felt that tylosin is particularly suitable as an anti-staphylococcal agent since the antibiotic possesses a narrow antibacterial spectrum and thus cannot be substituted for good sanitary practices. Furthermore, it was believed that tylosin would not affect the intestinal flora of persons who consumed food fortified with the antibiotic. Seitz *et al.* (1963) found tylosin to be effective in controlling growth of milk-borne staphylococci. This antibiotic also retarded acid production by starter cultures when it was present at concentrations in excess of 0.25 p.p.m. Lactobacilli appeared to be less susceptible to tylosin than were certain lactic starter cultures.

The effect of subtilin and subtilin-oxytetracycline combinations on growth of food poisoning salmonellae was studied by Godkin and Cathcart (1953). Development of salmonellae in custard was retarded for at least 72 hours when an incubation of 98.6° F. was used and when the custard contained 100 p.p.m. of subtilin plus ten p.p.m. of oxytetracycline. Food poisoning enterococci were inhibited for 72 hours by 100 p.p.m. of subtilin and for six to seven days by the subtilin plus 1.0 p.p.m. of oxytetracycline.

Rehm (1960) exposed *E. coli* to the combined effects of several antibiotics. He concluded that a) tetracycline and chloramphenicol displayed both additive and antagonistic activity but never synergism, b) streptomycin displayed synergism when in combination with the propyl ester of *p*-hydroxybenzoic acid, boric acid, sorbic acid, and sodium formate, and c) the behavior of the four antibiotics (chlortetracycline, tetracycline, chloramphenicol, and streptomycin) in combination was similar to the behavior of antibiotics and other preservatives in combination.

2. Sporeforming bacteria. — Andersen and Michener (1950 a and b) observed that foods could be preserved by the combination of subtilin and a mild heat treatment. Tests by Burroughs and Wheaton (1951) indicated that use of subtilin at 80 p.p.m., gramicidin at 100 p.p.m., bacitracin at 40 p.p.m., or streptomycin at 40 p.p.m. in conjunction with boiling was inadequate to prevent spoilage of canned peas or corn through growth of *C. botulinum.* Similar observations were made by Williams and Campbell

(1951). LEBLANC *et al.* (1953) reported that spores of putrefactive anaerobe No. 3679 and of *C. botulinum* were appreciably less heat resistant when suspended in a food containing small amounts of subtilin than when the food was without the antibiotic. The decrease in heat resistance of *Clostridium* spp. and *C. botulinum* in the presence of subtilin and nisin was verified by MICHENER (1953 a and b) and O'BRIEN *et al.* (1956). The latter authors observed further that this phenomenon also occurred with spores of *B. coagulans, B. stearothermophilus,* and *Clostridium thermosaccharolyticum.* Later DENNY and BOHRER (1959) showed the thermal death time of spores of putrefactive anaerobe No. 3679 was unaffected by subtilin, gramicidin, or tyrothricin. The latter two antibiotics did, however, serve to reduce the thermal death time of *B. stearothermophilus* spores.

ANDERSEN (1952) suggested that outgrowth of *C. botulinum* spores was prevented by 0.4 p.p.m. of subtilin and reduced 50 percent by 0.1 p.p.m. Spores of *B. stearothermophilus* and *B. coagulans* were themselves unaffected by the antibiotic, but 0.1 p.p.m. of subtilin prevented their growth (ANDERSEN and MICHENER 1950 a). MICHENER (1953 b) found *B. stearothermophilus* produced rough and smooth colonies; spores of the latter were unaffected by subtilin while those of the former produced subtilin-sensitive vegetative cells. According to later work by ANDERSEN *et al.* (1953) outgrowth of *C. botulinum* spores, as evidenced by colony formation, was prevented by 0.2 p.p.m. of penicillin or oxytetracycline, 1.0 p.p.m. of subtilin, 5.0 p.p.m. of chloramphenicol, gramicidin, or cinnamycin, and 20 p.p.m. of chlortetracycline, laterosporin, a different cinnamycin preparation, actinomycin, thiolutin, or nisin.

MICHENER (1955) observed that subjecting spores of *B. subtilis* to mild heat treatments (85° or 100° C.) made them more sensitive to the action of subtilin. He suggested that the antibiotic was adsorbed by the ungerminated spores and did not affect them until they were placed in an environment suitable for germination. Trypsin was able to inactivate the adsorbed subtilin. A similar increase in sensitivity after heating was noted by O'BRIEN and TITUS (1955) with putrefactive anaerobe No. 3679.

Tylosin and nisin were evaluated for their ability to inhibit food spoilage organisms by DENNY *et al.* (1961); tylosin displayed strong inhibitory activity against a butyric anaerobe, *B. coagulans,* and *B. stearothermophilus,* was markedly heat stable, and exhibited more activity than did nisin. According to GREENBERG and SILLIKER (1962 b), tylosin at levels up to 100 p.p.m. failed to inhibit germination of spores of *B. cereus* and putrefactive anaerobe No. 3679. Both organisms were susceptible to the antibiotic at or prior to the start of exponential growth. POOLE and MALIN (1964) noted that spores derived from tylosin-resistant vegetative cells gave rise to tylosin-sensitive vegetative cells.

Tylosin appeared quite inhibitory to *C. botulinum* according to GREENBERG and SILLIKER (1962 a). Meat products inoculated with *C. botulinum* and fortified with five, ten, 20, and 25 p.p.m. of tylosin were pasteurized at 68 to 74° C. After an appropriate incubation botulinus toxin was recovered from all foods prepared without tylosin but was absent from the tylosin-treated products. Viable botulinus organisms were also absent from the treated foods. Spoilage of the foods by non-botulinus organisms was encountered in many instances. MALIN *et al.* (1964) found a combination of 2.5 p.p.m. of tylosin and three percent sodium chloride prevented outgrowth of *C. botulinum* spores (11 strains) in fluid thioglycollate and trypticase-peptone-glucose-thioglycollate media.

Acknowledgement

The author acknowledges with thanks the assistance of Miss MYRTLE MAINA, Mrs. NANCY DENK, and Miss HELEN BROWNE in assembling references used in the preparation of this paper.

Summary

Antimicrobial substances in foods may originate from three principal sources. They may occur naturally, be produced during processing, or may be added intentionally or as contaminants.

Naturally occurring inhibitory substances have been recovered from a wide variety of foods including onion, garlic, horseradish, banana, sweet potato, cabbage, radish, fruits, cereals, spices, essential oils, egg white, milk, honey, hops, cod liver oil, and others. Some of the compounds found to occur naturally have been extracted, purified, and found useful for the preservation of some foods. Many of them undoubtedly contribute to the foodstuff's natural resistance to deterioration.

Antibiotic substances may be developed when some foods undergo a fermentation during processing. Homo- and heterofermentative streptococci, lactobacilli, propionibacteria, coliforms, *Bacillus* spp., and *Acetobacter* spp. are some of the microorganisms which appear in food fermentations and may elaborate inhibitory materials. Nisin, an antibiotic produced by *Streptococcus lactis*, has been manufactured and utilized commercially for the preservation of some foods. Inhibitory powers have also been associated with some cultured dairy products such as yogurt, acidophilus milk, certain cheeses, and others.

Finally, antibiotics may appear in foods to which they have been added intentionally for purposes of preservation or in which they appear as chance contaminants. Addition of antibiotics has been suggested to enhance the shelf-life of human (mother's) milk, vegetables, fruits, poultry tissue, eggs, pork, beef, lamb, meat products, fish of many types, and milk. Antimicrobial substances may appear in poultry meat, eggs, milk, milk products, and human milk because the chemical has been employed in treatment of disease or incorporated into the diet. Addition of antibiotics to fermentation mashes has been found beneficial for control of microbial contaminants and may lead to residues in certain fermented beverages.

Tetracycline-type antibiotics, tylosin, and subtilin, suggested as food additives, control growth of food-poisoning bacteria such as staphylococci, salmonellae, and enterococci. The presence of subtilin in a food appears to increase sensitivity of spores to thermal destruction while tylosin inhibits growth and toxin production by botulinus organisms.

Résumé *

Les substances antimicrobiennes dans les aliments peuvent provenir de trois sources principales. Elles peuvent être recontrées naturellement, être produites pendant la préparation ou être incorporées intentionnellement ou accidentellement par suite de contaminations.

Des substances inhibitrices recontrées naturellement ont été retrouvées dans une grande variété d'aliments comme l'oignon, l'ail, le raifort, la banane, la patate douce, le choux, les radis, les fruits, les céréales, les épices, les huiles essentielles, le blanc d'oeuf, le lait, le miel, le houblon, l'huile de foie de morue et d'autres. Quelques uns des composés recontrés naturellement ont été extraits, purifiés et reconnus utiles pour la conservation de quelques aliments. Beaucoup d'entre eux contribuent sans aucun doute à la résistance naturelle à la détérioration de ces aliments.

Quelques substances antibiotiques peuvent être produites lorsque certains aliments subissent une fermentation pendant leur préparation. Des streptocoques homo-et hétérofermentatifs, des lactobacilles, des bactéries propioniques, des Coliformes, le *Bacillus* spl et l'*Acétobacter* spl sont parmi les microorganismes qui apparaissent pendant la fermentation d'aliments et qui peuvent élaborer des éléments inhibiteurs. La Nisine, un antibiotique produit par *Streptococcus lactis,* a été manufacturé et utilisé commercialement pour la conservation de quelques aliments. Des facteurs inhibiteurs ont également été associés avec quelques produits diététiques tels que les yoghourts, le lait à l'acidophilus, certains fromages et d'autres denrées.

Enfin, les antibiotiques peuvent être trouvés dans des aliments auxquels ils sont été ajoutés intentionnellement à des fins de conservation ou dans lesquels ils apparaissent comme contaminants accidentels. L'addition d'antibiotiques a été suggérée pour augmenter la durée de conservation du lait de femme, des légumes, fruits, chair de volailles, oeufs, porc, boeuf, agneau, viandes, poissons variés et lait. Des substances antimicrobiennes peuvent apparaître dans la chair de volaille, les oeufs, le lait, les produits laitiers et le lait humain, lorsqu'un produit chimique a été employé dans le traitement de maladies ou incorporé au régime. L'addition d'antibiotiques à des bouillies de fermentation à été reconnue bénéfique pour lutter contre les contaminations microbiennes et peut conduire à des résidus dans certaines boissons fermentées. Les antibiotiques du type tetracycline, la tylosine et la subtiline, ont été proposés comme additifs alimentaires, pour inhiber la croissance de microorganismes pouvant contaminer les aliments comme les staphylocoques, salmonella et entérocoques. La présence de subtiline dans un aliment semble augmenter la sensibilité des spores à la destruction thermique alors que la tylosine inhibe la croissance et la production de toxine par les microbes du botulisme.

Zusammenfassung **

Antimikrobielle Substanzen in Nahrungsmitteln können von 3 Hauptquellen stammen: sie können natürlich vorkommen, während der Verarbeitung erzeugt werden oder absichtlich als Verunreinigung zugefügt sein.

* Traduit par R. TRUHAUT.
** Übersetzt von H. MARTIN.

Natürlich vorkommende Hemmsubstanzen wurden von einer großen Zahl Nahrungsmitteln gewonnen. Zu ihnen gehören Zwiebel, Knoblauch, Meerrettich, Banane, Süßkartoffel, Kohl, Rettich, Obst, Getreide, Gewürze, ätherische Öle, Eiweiß, Milch, Honig, Hopfen, Lebertran u. a. Einige der natürlich vorkommenden Verbindungen wurden extrahiert und gereinigt und zur Nahrungsmittel-Konservierung geeignet gefunden. Zweifellos tragen viele von ihnen zur natürlichen Widerstandskraft gegen Fäulnis bei.

Antibiotische Substanzen können erzeugt werden, wenn einige Nahrungsmittel während des Herstellungsprozesses gären. Selbst- und fremdgärende Streptokokken, Laktobazillen, Propionbakterien, Koliformen, *Bacillus* sp. und *Acetobacter* sp. sind einige der Mikroorganismen, die bei der Gärung von Nahrungsmitteln auftreten und Hemmstoffe hervorrufen können. Nisin, ein Antibioticum, das vom *Streptococcus lactis* erzeugt wird, wurde hergestellt und kommerziell zur Nahrungsmittelkonservierung benutzt. Einigen Erzeugnissen der Milchverarbeitung, wie Yoghurt, azidophile Milch, gewisse Käsesorten, können Hemmwirkungen zugeschrieben werden.

Schließlich erscheinen Antibiotica auch in Nahrungsmitteln, denen sie absichtlich zur Konservierung beigefügt wurden oder in denen sie als zufällige Beimischungen erscheinen. Zusatz von Antibiotica zur Verlängerung der verkaufsfähigen Spanne wurde für Muttermilch, Gemüse, Obst, Geflügel, Eier, Schweinefleisch, Rindfleisch, Lamm, weitere Fleischsorten, viele Arten von Fisch und Milch vorgeschlagen. Antimikrobielle Substanzen können in Geflügelfleisch, Eiern, Milch und Milchprodukten und in Muttermilch auftreten, wenn die chemische Substanz zur Behandlung von Krankheit benutzt oder der Nahrung zugesetzt wurde. Der Zusatz von Antibiotica hat sich als günstig für die Kontrolle mikrobieller Ansteckungskeime in gärenden Maischen erwiesen und kann daher zu Rückständen in gewissen vergorenen Getränken führen.

Antibiotica vom Typ des Tetrazyklin, Tylosin und Subtilin, die als Nahrungsmittel-Zusätze vorgeschlagen werden, verhindern das Wachstum von nahrungsmittelvergiftenden Bakterien wie Staphylokokken, Salmonellen und Enterokokken. Die Anwesenheit von Subtilin in Nahrungsmitteln scheint die Empfindlichkeit von Sporen gegenüber thermischer Zerstörung zu erhöhen, während Tylosin Wachstum und Toxinproduktion von Botulinus-Arten hemmt.

References

Akiyama, H., K. Yamao, and S. Sasaki: Antibiotics for controlling contamination in sake brewing. Nippon Jôzô Kyôkai Zasshi **53**, 306 (1958).

Albright, J. L., S. L. Tuckey, and G. T. Woods: Antibiotics in milk — a review. J. Dairy Sci. **44**, 779 (1961).

Alifax, R., and R. Chevalier: Nisinase produced by *Streptococcus thermophilus*. J. Dairy Research **29**, 233 (1962).

Allen, P. T., and H. Beckman: Polarography for the determination of organic feed medicaments. Residue Reviews **5**, 91 (1964).

Andersen, A. A.: Effect of subtilin on spores of *Clostridium botulinum*. J. Bacteriol. **64**, 145 (1952).

—, and H. D. Michener: Effect of subtilin on bacterial spores. Bact. Proc. **1950**, 28 (1950 a).

ANDERSEN, A. A., and H. D. MICHENER: Preservation of foods with antibiotics. I. The complementary action of subtilin and mild heat. Food Technol. **4**, 188 (1950 b.

— —, and H. S. OLCOTT: Effect of some antibiotics on *Clostridium botulinum.* Antibiotics and Chemotherapy **3**, 521 (1953).

ANONYMOUS: Antibiotic-resistant organisms in cheese. J. Amer. Med. Assoc. **159**, 1788 (1955).

— Penicillin and other antibiotics in milk. J. Amer. Med. Assoc. **171**, 135 (1959 a).

— Penicillin and other antibiotics in milk. J. Milk Food Technol. **22**, 344 (1959 b).

ARK, P. A., and J. P. THOMPSON: Antibiotic properties of the seeds of wheat and barley. Plant Disease Reptr. **42**, 959 (1958).

— — Control of certain diseases of plants with antibiotics from garlic *(Allium sativum* L.). Plant Disease Reptr. **43**, (1959).

ARPAI, J., M. BEHUN, Z. LIFKOVA, and D. VRABLICOVA: The influence of additional feeding of pigs with Aureovit 12 on the quality of meat with reference to its preservation by freezing. Sb. Cesk. Akad. Zimedel. Ved, Zivocisna Vyroba **5**, 725 (1959).

ASSELBERGS, E. A. M., W. E. FERGUSON, W. P. MOHR, and K. F. MACQUEEN: Effects of sodium metabisulfite, antibiotics and gamma irradiation on the shelflife of prepeeled potatoes at room temperature. Applied Microbiol. **8**, 311 (1960).

AUCLAIR, J.: The inhibition by penicillin of lactic acid starters used in the manufacture of Gruyère cheese. Lait **31**, 121 (1952).

AYRES, J. C.: Use of coating materials or film impregnated with chlortetracycline to enhance color and storage life of fresh beef. Food Technol. **13**, 512 (1959).

— Effect of sanitation, packaging, and antibiotics on microbial spoilage of commercially processed poultry. Conference on eggs and poultry. U.S. Dept. Agr., Agr. Research Service Rept. ARS-74-12, p. 18 (1959).

—, and E. L. DENISEN: Maintaining freshness of berries using selected packaging materials and antifungal agents. Food Technol. **12**, 562 (1958).

BABAD, J., D. L. BOROS, and F. BAIER: Lowered bactericidal efficiency of hydrogen peroxide on milk from cows treated with penicillin. Nature **184**, 292 (1959).

BAKER, E. G., B. A. SOUTHCOTT, and H. L. A. TARR: Effect of chlortetracycline (CTC) antibiotic on the keeping quality of lingcod stored in refrigerated sea water. Progress Rept. Pacific Coast Stations Fisheries Research Board of Canada, No. **112**, pp. 15—17 (1958).

BARBER, R. S., R. BRAUDE, and A. HIRSCH: Growth of pigs given skim milk soured with nisin-producing streptococci. Nature **169**, 200 (1952).

BARNES, E. M., and D. H. SHRIMPTON: The effect of the tetracycline compounds on the storage life and microbiology of chilled eviscerated poultry. J. Appl. Bacteriol. **21**, 313 (1958).

BARTA, J.: Antibiotics in the yeast industry. Kvasny Prumysl **6**, 283 (1960).

BARTUSKOVA-CERNIKOVA, M.: The use of nisin for controlling the blowing of processed cheese. Sb. Vysoke Skoly Chem.-Tech. Praze. Tech. **6**, 330 (1962).

BAUGHMAN, R. W., and F. E. NELSON: Effect of penicillin upon the morphology and gram-staining characteristics of *Streptococcus lactis* and *Streptococcus thermophilus.* J. Dairy Sci. **41**, 706 (1958).

BAUMGARTNER, H.: Die Penicillinbehandlung des Euter. Schweiz. Arch. Tierheilk. **89**, 215 (1947).

BEACH, A. S.: An agar diffusion method for the assay of nisin. J. Gen. Microbiol. **6**, 60 (1952).

BENARDE, M. A.: Antibiotic residues in shellfish after cooking. J. Amer. Dietetic Assoc. **33**, 1145 (1957).

— Comparison of tap and distilled water antibiotic dip solutions on storage life of fresh crab meat. Antibiotic Ann. **1957—1958**, 224 (1958).

—, and R. A. LITTLEFORD: Antibiotic treatment of crab and oyster meats. Applied Microbiol. **5**, 368 (1957).

BERNHARD, E.: Action of metabolic products of *Bacillus laterosporus* Lauback on *Bacillus cereus* Frankland et Frankland. Prom. No. 3474 (1964).

BERRIDGE, N. J.: Preparation of the antibiotic nisin. Biochem. J. **45**, 486 (1949).

BERRIDGE, N. J.: The antibiotic nisin and its use in the making and processing of cheese. Chem. Ind. **1953**, 1158 (1953).
— Inhibitory substances of bacterial and other origins in milk and milk products. J. Sci. Food Agr. **6**, 65 (1955).
— Penicillin in milk. III. Effect of low concentrations of penicillin on the rate of acid production by starter cultures. J. Dairy Research **23**, 348 (1956).
—, and J. BARRETT: Rapid turbidimetric method for the assay of antibiotics. Nature **167**, 448 (1951).
— — A rapid method for the turbidimetric assay of antibiotics. J. Gen. Microbiol. **6**, 14 (1952).
—, G. G. F. NEWTON, and E. P. ABRAHAM: Purification and nature of the antibiotic nisin. Biochem. J. **52**, 529 (1952).
BINDER, W.: The formation of an antibiotic substance in moulds of the Camembert, Gorgonzola and Roquefort types of cheese. Milchwissensch. Ber. **4**, 88 (1954).
BINDEWALD, H.: Preservation of mothers' milk with streptomycin or citric acid. Deut. Med. Wochschr. **77**, 1015 (1952).
BIRD, H. R.: Additives and residues in food of animal origin. In: Symposium on additives and residues in human food. Amer. J. Clinical Nutrit. **9**, 260 (1961).
BISSETT, H. M., and H. L. A. TARR: Stability of antibiotics when used in experimentally retarding fish spoilage. Fisheries Research Board of Canada, Progress Rept. Pacific Coast Station, No. **93**, 23 (1952).
BLOOM, W. L., and J. R. PRIGMORE: A method for the preparation of antibacterial basic proteins of normal tissues. J. Bacteriol. **64**, 855 (1952).
BLUHM, H. M., and H. L. A. TARR: Effect of chlortetracycline treatment of fish on development of certain food-poisoning bacteria. Progress Rept. Pacific Coast Stations Fisheries Research Board of Canada, No. **106**, pp. 8—10, (1956).
BLUM, S., F. NOVAK, and S. FABER, III: 10-Hydroxy-Δ2-decenoic acid, an antibiotic found in royal jelly. Science **130**, 452 (1959).
BOCKELMANN, J. B., and F. B. STRANDSKOV: Inhibition of microbiological growth in beer. U.S. Pat. No. 2,798,811 (1957).
BONDARENKO, A. S., and S. I. ZELEPUKHA: Antibacterial properties of strawberries. Mikrobiol. Zh., Akad. Nauk Ukr. RSR **24**, 41 (1962).
BONDE, R.: Preliminary studies on the control of bacterial decay of the potato with antibiotics. Amer. Potato J. **30**, 143 (1953).
BORRIE, P., and J. BARRETT: Dermatitis caused by penicillin in bulked milk supplies. Brit. Med. J. **1961-II**, 1267 (1961).
BOYD, J. W., and B. A. SOUTHCOTT: Tetracycline antibiotics in objective and subjective fish quality tests. J. Fisheries Research Board Canada **19**, 615 (1962).
—, and H. L. A. TARR: Effect of chlortetracycline and storage temperatures on the quality of shucked oysters. Progress Rept. Pacific Coast Stations Fisheries Research Board of Canada, No. **105**, pp. 12—13 (1956).
—, H. M. BISSETT, and H. L. A. TARR: Effect of addition of aureomycin to sea water on the viable bacteria present in the viscera of living fish. Progress Rept. Pacific Coast Stations Fisheries Research Board of Canada, No. **102**, pp. 4—5 (1955 a).
— — — Further observations on the distribution of chlortetracycline throughout ice blocks. Progress Rept. Pacific Coast Stations Fisheries Research Board of Canada, No. **102**, pp. 14—15 (1955 b).
—, C. BRUNWELL, and H. L. A. TARR: Aureomycin in experimental fish preservation. Progress Rept. Pacific Coast Stations Fisheries Research Board of Canada, No. **96**, p. 25 (1953).
—, B. A. SOUTHCOTT, and H. L. A. TARR: Antibiotic residues in fish iced with chlortetracycline ice and effect of normal cooking procedures on these residues. Antibiot. Ann. **1956—1957**, 1002 (1957).
—, H. M. BLUHM, C. R. MUIRHEAD, and H. L. A. TARR: Use of antibiotics for the preservation of fish and sea food. Amer. J. Public Health **46**, 1531 (1956).
BOYD, JOSEPHINE, H. H. WEISER, and A. R. WINTER: Influence of high level antibiotic rations with terephthalic acid on antibiotic content and keeping quality of poultry meat. Poultry Sci. **39**, 1067 (1960).

BRADFIELD, A.: Effect of mastitis curatives on cheese making. Proc. 28th. Ed. Conf. Vermont Dairy Plant Operators and Milk Distributors, pp. 23—25 (1949).

— The effects of mastitis curatives on cheese making. Can. J. Compar. Med. Vet. Sci. **14**, 127 (1950).

BREDENBERG, J. B., and P. K. HIETALA: Confirmation of the structure of trifolirhizin. Acta Chim. Scand. **15**, 936 (1961 a).

— — Investigation of the structure of trifolirhizin, an antifungal compound from *Trifolium pratense* L. Acta Chim. Scand. **15**, 696 (1961 b).

BROQUIST, H. P., A. R. KOHLER, and W. H. MILLER: Retardation of poultry spoilage by processing with chlortetracycline. J. Agr. Food Chem. **4**, 1030 (1956).

BRUCKNER, B. H., H. McKAY, P. S. SHAFFER, and T. D. FONTAINE: The partial purification and properties of antibiotic substances from the sweet potato plant *(Ipomoea batatas)*. J. Clinical Invest. **28**, 894 (1949).

BRÜGGEMANN, J., H. KARG, and K. BARTH: Reaction of albino rats to chronic feeding with chicken meat pretreated with chlortetracycline. Studies on generations of rats. Arzneimittel-Forsch. **11**, 1022 (1961).

—, M. MERKENSCHLAGER, and K. H. NIESAR: Bilanzstudien mit ^{14}C-Chlortetracyclin an der Legehenne unter den Bedingungen der Antibiotikafütterung. Zentr. Veterinarmed. **10**, 631 (1963).

BRYAN, C. S.: Problems created for the dairy industry by antibiotic mastitis treatments. J. Milk Food Technol. **14**, 161 (1951).

BUKANOVA, V. I.: Antibiotic properties of some sour-milk products. Gigiena i Sanit. **1952**, 32 (1952).

BURROUGHS, J. D., and I. E. WHEATON: The preservative action of antibiotics in processed foods. Canner **112**, 50 (1951).

BUSTINZA, F.: On the antibiotic activity of the juice of the bulb of *Allium cepa* L. Ann. Inst. Bot. (Madrid) **17**, 3 (1959).

BYLUND, G., O. NORRGREN, and G. SJOSTROM: The use of antibiotic-producing strains of *Streptococcus lactis* to combat butyric acid fermentation in cheese. An orientating experiment. Svenska Mejeritidn. **46**, 433 (1954).

BYRNES, J. M., and R. COREY: Oxytetracycline resistant bacteria in commercial poultry products. Bacteriol. Proc. **1961**, 70 (1961).

CAIN, R. F., A. W. ANDERSON, and A. S. MALASPINA: Effect of radiation on antibiotic-treated meats. Food Technol. **12**, 582 (1958).

CAMERON, E. J.: Use of antibiotics in preserving foods. Proc. 3rd. Conf. on Research, Council on Research, Amer. Meat Inst. (1951).

CAMPBELL, L. L. Jr., and E. E. SNIFF: Nisin sensitivity of *Bacillus coagulans*. Applied Microbiol. **7**, 289 (1959).

— —, and R. T. O'BRIEN: Subtilin and nisin as additives that lower the heat-process requirements of canned foods. Food Technol. **13**, 462 (1959).

CAPEK, A.: Phytoncides of anise and fennel. Prumysl Potravin **7**, 260 (1956).

— Phytoncides of spices. Prumysl Potravin **6**, 433 (1955).

CAREY, B. W.: The biochemistry of antibiotics on the bacteria associated with food spoilage. Giorn. Microbiol. **2**, 233 (1956).

CARLSON, S., and H. M. BAUER: Nisin, and antibacterial substances from *Streptococcus lactis*, with consideration of the problem of resistance. Arch. Hyg. Bacteriol. **141**, 445 (1957).

CARROLL, V. J., R. A. BENEDICT, and C. L. WRENSHALL: Delaying vegetable spoilage with antibiotics. Food Technol. **11**, 490 (1957).

CASTELL, C. H.: The keeping time of wrapped fillets treated with antibiotic. Canadian Fisherman **45**, 6 (1958).

— Chlorine affects antibiotics. Canadian Fisherman **47**, 37 (1960).

—, and J. DALE: Antibiotic dips for preserving fish fillets. Bull. Fisheries Research Board No. **138**, 70 pp. (1963).

—, and M. F. GREENOUGH: Effect of certain antibiotics on the production of trimethylamine and hydrogen sulfide by bacterial enzymes. J. Fisheries Research Board Canada **14**, 771 (1957).

CASTELL, C. H., and M. F. GREENOUGH: The use of antibiotics for the control of spoilage in the East Coast fisheries. I. Introduction and the use of antibiotics in gutted cod and haddock aboard trawlers. Canadian Fisherman **45**, 6 (1958 a).
— — The use of antibiotics for the control of spoilage in the East Coast fisheries. II. The effect of antibiotics on the keeping time of fillets. Canadian Fisherman **45**, 20 (1958 b).
CATES, F. B., and L. O. VAN BLARICOM: The effect of several compounds on post-harvest decay of potatoes. Amer. Potato J. **38**, 175 (1961).
CAVALLITO, C. J.: Extraction of garlic. U.S. Pat. No. 2,554,088 (1951).
CHANDAN, R. C., and K. M. SHAHANI: The milk and microbial lipases as affected by antibiotics. J. Dairy Sci. **43**, 841 (1960).
—, M. G. CARRANCEDO, and K. M. SHAHANI: A study of the inhibition of microbial lipase by antibiotics. J. Dairy Sci. **44**, 1161 (1961).
CHEESEMAN, G. C., and N. J. BERRIDGE: Resistance of nisin to penicillinase. Acta Chim. Scand. **8**, 127 (1954).
— — Improved method of preparing nisin. Biochem. J. **65**, 603 (1957).
— — Molecular weight and chemical composition of nisin *A*. Biochem. J. **71**, 185 (1959).
CHEVALIER, R., and G. MOCQUOT: A nisin titration method using the clotting of milk with rennet. 15th Internat. Dairy Congress, London **3**, 1436 (1959).
CHURCHILL, E. S., A. R. DRURY, B. D. LEWIS, C. L. FRANK, and C. S. BRYAN: A report on the modified phosphatase method of detecting antibiotic substances in milk. Milk Plant Monthly **40**, 28 (1951).
CHUZHOVA, Z.: Medicinal sour milk beverage "Kurunga" and its antibiotic properties. Molochnaya. Prom. **19**, 34 (1958).
COLLINS, E. B.: Domination among strains of lactic streptococci with attention to antibiotic production. Applied Microbiol. **9**, 200 (1961).
COOPER, G. M., and D. K. SALUNKHE: Effect of gamma-radiation, chemical and packaging treatments on refrigerated life of strawberries and sweet cherries. Food Technol. **17**, 123 (1963).
COREY, R. R., and J. M. BYRNES: Oxytetracycline-resistant coliforms in commercial poultry products. Applied Microbiol. **11**, 481 (1963).
COTTERILL, J., and P. HARTMAN: Effect of antibiotics on the incidence of spoilage of shell eggs. Poultry Sci. **35**, 733 (1956).
CREAN, P. B., H. L. A. TARR, and R. B. BARKER: Control of *post-mortem* bacterial spoilage of whales with chlortetracycline. Progress Rept. Pacific Coast Stations Fisheries Research Board of Canada, No. **105**, pp. 6—10 (1956).
CRUICKSHANK, I. A., and D. R. PERRIN: Isolation of a phytoalexin from *Pissum sativum*. Nature **187**, 799 (1960).
CSISZAR, J., and G. PULAY: Research on *Streptococcus lactis* producing antibiotic against clostridia. 14th Internat. Dairy Congress, Rome **2**, 423 (1956).
CZETSCH-LINDENWALD: Antibiotic medicinal plants and their possible uses. Planta Med. **4**, 209 (1956).
DAL-CIN, G.: Antibiotics in wine making. Riv. viticolt. e enol. (Coneglians) **1**, 335 (1948).
DAMEROW, R.: A criticism of human milk preservation with streptomycin. Arch. Kinderheilk. **152**, 56 (1956).
DANNON, S., S. ZHEKOV, and M. KOZAREVA: Role of lactic acid in the antibiotic effect of yoghurt. C. R. Acad. Bulgar. Sci. **13**, 749 (1960).
DATTA, N. L., A. KRISHNAMURTHI, and S. SIDDIQUI: Antibiotic principles of *Allium sativum* (clove). J. Sci. Ind. Research **7 B**, 42 (1948).
DAWSON, L. E., and W. J. STADELMAN: Microorganisms and their control on fresh poultry meat. Mich. State Univ., Tech. Bull. No. 278, pp. 1—40 (1960).
—, W. L. MALLMANN, D. G. BIGBEE, R. WALKER, and M. E. ZABIK: Influence of surface pasteurization and chlortetracycline on bacterial incidence on fryers. Food Technol. **17**, 218 (1963).
DAY, W. H., W. C. SERJAK, J. R. STRATTON, and L. STONE: Antibiotics as contamination control agents in grain alcohol fermentations. J. Agr. Food Chem. **2**, 252 (1954).
DEATHERAGE, F. E.: Preservation of meat. U. S. Pat. No. 2,786,768 (1956).

DEATHERAGE, F. E.: Use of antibiotics in the preservation of meats and other food products. Amer. J. Public Health **47**, 594 (1957).
DE CLEREK, J.: Antibiotics of hops and beer. Symp. Substances Etranger dans les Aliments, 6e, Madrid **1960**, 254 (1961).
DE KLERK, H. C., and J. N. COETZEE: Antibiosis among lactobacilli. Nature **192**, 340 (1961).
DENNY, C. B., and C. W. BOHRER: Effect of antibiotics on the thermal death rate of spores of food spoilage organisms. Food Research **24**, 247 (1959).
—, J. M. REED, and C. W. BOHRER: Effect of tylosin and heat on spoilage bacteria in canned corn and mushrooms. Food Technol. **15**, 338 (1961).
—, L. E. SHARPE, and C. W. BOHRER: Effects of tylosin and nisin on canned food spoilage bacteria. Applied Microbiol. **9**, 108 (1961).
DE SILVA, N. N., and R. B. HUGHES: Effect of tetracycline antibiotics on the proteolysis of fish muscle. Nature **188**, 663 (1960).
— — Influence of tetracycline antibiotics on the spoilage of herring. J. Sci. Food Agr. **13**, 161 (1962).
DEVYATKINA, O. V., M. P. SEDUNOV, and I. V. KHUZINA: Treatment for storage of fresh fish in aqueous biomycin. Nauchn. Tekhn. Byul, Polyarnyi Nauchn Issled Inst. Morsk, Rybn. Khoz. i Okeanogr. **1961**, 33 (1961).
DICKERMAN, J. M., and S. LIBERMAN: Chemical nature of an antibiotic present in water extracts of cabbage. Food Research **17**, 438 (1952).
DICKINSON, L., and C. F. POE: Preservation of milk with chlortetracycline. Univ. Colo. Stud. Series Chem. and Pharm. **2**, 50 (1959).
DO, J. Y., and D. K. SALUNKHE: Effect of chemical treatments on refrigerated apple juice. Food Technol. **18**, 182 (1964).
DOLD, H., and R. WITZENHAUSEN: Ein Verfahren zur Beurteilung der örtlichen inhibitorischen (keimvermehrungshemmenden) Wirkung von Honigsorten verschiedener Herkunft. Hyg. Infektionskrankh. **141**, 333 (1955).
—, D. H. DU, and S. T. DZIAO: Demonstration of a heat and light sensitive inhibitory substance (inhibine) in natural honey. Hyg. Infektionskrankh. **120**, 155 (1937).
DUBINSKII, R. A.: Relation between the antimicrobial activity and the acidity of fermented milk products. Vopr. Pitaniya **17**, 68 (1958).
DUBROVA, G. B.: Use of phytoncide and antibiotic combinations for preservation of meat and fish. Sb. Nauch. Rabot. Leningr. Inst. Sovet. Torgovli **1955**, 139 (1955).
— Use of phytoncides for conservation of foodstuffs. Tr. Leningrad Tekhnol. Inst. Kholodil. Prom. **10**, 74 (1956).
DUISBERG, H., und B. WARNECKE: Erhitzungs- und Lichteinfluß auf Fermente und Inhibine des Honigs. Z. Lebensm. Untersuch. u. Forsch. **111**, 111 (1959).
DUMAIS, R.: L'effect nuisible de la penicilline dans ce lait sur les ferments lactiques et le fromage chèddar. Lait **10**, 14 (1951).
DUNCAN, D. W., J. LEUNG, N. A. W. LE PAGE, J. W. BOYD, and H. L. A. TARR: Control of *post-mortem* bacterial spoilage of whales with tetracycline antibiotics. Progress Rept. Pacific Coast Stations Fisheries Research Board of Canada, No. 109, pp. 14—15 (1957).
EASTOE, J. E., and J. E. LONG: The effect of nisin on the growth of cells and spores of *Clostridium welchii (C. perfringens)* in gelatin. J. Appl. Bacteriol. **22**, 1 (1959).
EKLUND, M. W., J. V. SPENCER, E. A. SAUTER, and M. H. GEORGE: Effect of different methods of chlortetracycline application on shelf life of chicken fryers. Poultry Sci. **40**, 924 (1961).
ELLIOT, L. E., and G. L. ROMOSER: Recovery of antibiotic residue in egg albumin. Poultry Sci. **36**, 365 (1957).
EMANUILOFF, I., and L. NATSCHEFF: Research into the antibiotic activity of the microflora of yoghurt. 14th Internat. Dairy Congress, Rome **1**, 548 (1956).
EMMERIE, A., C. ENGEL, and W. KLIP: Bacteriostatic substances in the unsaponifiable fraction of cod liver oil. Nature **166**, 787 (1950).

EOFF, H. J., R. E. DAVIES, T. M. FERGUSON, and J. R. COUCH: Influence of terephthalic acid and broad-spectrum antibiotics on egg production and egg shell coloration in caged layers. Poultry Sci. **41**, 1036 (1962).

ERSKINE, D.: Dermatitis caused by penicillin in milk. Lancet **1**, 431 (1958).

ESSARY, E. O., L. E. DAWSON, and W. L. MALLMANN: Effect of different length chill periods with chlortetracycline and different holding conditions on shelf life of dressed fryers. Food Technol. **14**, 286 (1960).

—, E. C. MOORE, and C. Y. KRAMER: Influence of scald temperatures, chill time, and holding temperatures on the bacterial flora and shelf-life of freshly chilled, tray-pack poultry. Food Technol. **12**, 684 (1958).

EVANS, F. R., and H. R. CURRAN: The preserving action of subtilin and mild heat in normal and concentrated milk. J. Dairy Sci. **35**, 1101 (1952).

EZDAKOV, N. V.: The fattening of pigs with residues of the antibiotic industry. Primenenie Antibiotikov v Zhivotnovodstve, Materialy Soveshch. Otd. Zhivotnovodstva, Vses. Akad. Sel'skokhoz. Nauk im. V. I. Lenina Moscow **1958**, 48 (1959).

FALCONER, R.: Antibiotic substances produced by *Streptococcus lactis*. Brit. Pat. No. 683, 423 (1952).

FARBER, L.: Antibiotics as aids in fish preservation. I. Studies on fish fillets and shrimp. Food Technol. **8**, 503 (1954).

—, and P. LERKE: Plant-scale applications of chlortetracycline for fish preservation. Univ. Miami Lab. Proc. Gulf Carribbean Fisheries Inst. **10**, 72 (1957).

FAVILLE, L. W., and F. W. FABIAN: The influence of bacteriophage, antibiotics, and Eh on the lactic fermentation of cucumbers. Mich. Agr. Expt. Station Tech. Bull. No. 217 (1949).

FEENEY, R. E., and D. A. NAGY: The antibacterial activity of the egg white protein conalbumen. J. Bacteriol. **64**, 629 (1952).

FELDMAN, Z. Y.: Utilization of antibiotics in storing lemon. Tr. Leningr. Tekhnol. Inst. Kholodiln. Prom. **1956**, 72 (1956).

FIEGER, E. A., M. E. BAILEY, and A. F. NOVAK: Chemical ices for shrimp preservation. Food Technol. **10**, 578 (1956).

FIRMAN, M. C., A. ABBEY, M. A. DARKEN, A. R. KOHLER, and S. D. UPHAM: Effect of aureomycin-chlortetracycline on fish freshness. Food Technol. **10**, 381 (1956).

FISCHER, G.: A phagicidal and virucidal agent in maple fruit *(Acer platanoides)* preparations. Investigations with *Escherichia coli* and vaccinia virus. Acta Pathol. Microbiol. Scand. **31**, 433 (1952).

FOLEY, E. J., and J. V. BYRNE: Penicillin as an adjunct to the preservation of quality of raw and pasteurized milk. J. Milk Food Technol. **13**, 170 (1950).

FOSTER, J. W., and H. B. WOODRUFF: Microbiological aspects of penicillin. I. Methods of assay. J. Bacteriol. **46**, 187 (1943).

FRANC, Z., I. M. HAIS, and O. HORESOVSKY: Antituberculous factors of milk. Nature **182**, 884 (1958).

FRANCIS, F. J.: Discoloration and quality maintenance in cole slaw. J. Amer. Soc. Sci. **75**, 449 (1960).

FRAZER, A. C., M. SHARRAT, and J. R. HICKMAN: Biological effects of food additives. I. Nisin. J. Sci. Food Agr. **13**, 32 (1962).

FRIEDLANDER, S.: Penicillinase in the treatment of allergic reactions to penicillin. J. Allergy **30**, 181 (1959).

FRIEDMANN, R., and C. EPSTEIN: Assay of the antibiotic nisin by means of a reductase (resazurin) test. J. Gen. Microbiol. **5**, 830 (1951).

FRYE, G. R., H. H. WEISER, and A. R. WINTER: Influence of antibiotic feeding on the keeping quality and antibiotic content of poultry products. Poultry Sci. **37**, 633 (1958).

FULLER, J. E., and E. R. HIGGINS: Onion juice and bacterial growth. Food Research **5**, 503 (1940).

GALESLOOT, T. E.: Influence of penicillin upon the morphology of yoghurt- and starter bacteria. A rapid method to establish penicillin as the probable cause of yoghurt and starter failures. Netherlands Milk Dairy J. **10**, 64 (1956 a).

GALESLOOT, T. E.: Lactic acid bacteria which destroy the antibioticum (nisin) of *Streptococcus lactis*. Netherlands Milk Dairy J. **10**, 143 (1956 b).
— Nisin production by antibiotic starters. Netherlands Milk Dairy J. **11**, 74 (1957 a).
— The effect of nisin on the growth of bacteria connected or possibly connected with the bacterial processes in cheese and processed cheese. Netherlands Milk Dairy J. **11**, 58 (1957 b).
— The propagation of nisin-producing starters. 15th Internat. Dairy Congress, London **2**, 547 (1959).
—, and F. HASSING: The stability of the nisin content of Edam and Gouda cheese made from pasteurized milk with a nisin-producing starter. Netherlands Milk Dairy J. **18**, 177 (1964).
—, and J. W. PETTE: The estimation of the nisin content of antibiotic starters and cultures and of cheese made by means of antibiotic starters. Netherlands Milk Dairy J. **10**, 137 (1956).
— — The formation of normal eyes in Edam cheese made by nisin-producing starters. Netherlands Milk Dairy J. **11**, 144 (1957).
—, G. W. RAADSVELD, and J. W. PETTE: The influence of nisin-producing starters upon the ripening of Edam cheese. Netherlands Milk Dairy J. **11**, 152 (1957).
GALLO, P., and H. VALERI: Determination of the antibiotic of the pericarp of the avocado pear *(Persea americana)*. Review Med. Vet. Parasitol. **13**, 63 (1955).
— — The antibiotic activity of the seed of avocado pear *(Persea americana)*. Review Med. Vet. Parasitol. **12**, 125 (1954).
GARCIA, F. R.: The influence of nisin in the prevention of blowing in cheese. Anales Fac. Vet. Leon **5**, 171 (1959).
GARVIE, ELLEN I., and M. JEAN STONE: The associative growth of *Bacillus cereus* and *Streptococcus lactis* in milk. J. Dairy Research **20**, 29 (1953).
GIBBS, B. M., and A. HURST: Limitation of nisin as a preservative in non-dairy foods. 4th Internat. Symp. on Food Microbiol., Sweden, June 1—5, 1964.
GIBSHMAN, M. R., and N. N. BELOUSOVA: Selection of starters in the manufacture of Dutch-type cheese. Dokl. Vses. Konf. po Molochn. Delu. **1958**, 356 (1958).
GILLESPIE, D. C., H. M. BISSETT, J. W. BOYD, and H. L. A. TARR: Aureomycin in experimental fish preservation. II. Progress Rept. Fisheries Research Board of Canada Pacific Coast Station, No. 100, 12 (1954 a).
— — — — A method of facilitation of germicidal substances throughout ice blocks. Progress Rept. of the Fisheries Research Board of Canada Pacific Coast Station, No. 99, pp. 18—19 (1954 b).
—, J. W. BOYD, H. M. BISSETT, and H. L. A. TARR: Ices containing chlorotetracycline in experimental fish preservation. Food Technol. **9**, 296 (1955).
GINSBERG, A., M. REID, J. M. GRIEVE, and K. OGONOWSKI: Chlortetracycline as a preservative for fresh meat and poultry. Vet. Record **70**, 700 (1958).
GIOLITTI, G., and F. GIANELLI: Conservation of eggs by means of antibiotics. Ateneo Parmense **31** (Suppl. 1), 73 (1960).
GODKIN, W. J., and W. H. CATHCART: Effect of antibiotics in retarding the growth of *Micrococcus pyogenes* var. *aureus* in custard fillings. Food Technol. **6**, 224 (1952).
— — The complementary action of subtilin and terramycin in preserving custard fillings. Food Technol. **7**, 282 (1953).
GOLA, J., V. SINGER, and M. LUKESOVA: The use of chlortetracycline (aureomycin) in prolonging the keeping quality of transported veal. Sb. Cek. Akad. Zemedel. Ved. Vet. Med. **5**, 495 (1959).
GOLDBERG, H. S.: Studies on the microflora and deep tissues of beef and their susceptibility to various antibiotics. Bacteriol. Proc. **1952**, 17 (1952).
—, H. H. WEISER, and F. E. DEATHERAGE: Studies on meat. IV. The use of antibiotics in preservation of fresh beef. Food Technol. **7**, 165 (1953).
GORNER, F., and L. KABATOVA: Nisin as a preventive agent of microbial alteration in processed cheese. Prumysl. Potravin **14**, 549 (1963).
—, and H. SIMKOVICOVA: Influence of nisin on microbial changes in processed cheese. Prumysl. Potravin **12**, 570 (1961).

GOWANS, J. L., N. SMITH, and H. W. FLOREY: Some properties of nisin. Brit. J. Pharmacol. **7**, 438 (1952).

GRAY, P. P., and A. D. KAZIN: Antibiotics and the treatment of brewer's yeast. Wallerstein Labs. Commun. **9**, 115 (1946).

GRECZ, N., G. M. DACK, and L. R. HEDRICK: Antimicrobial agent of aged surface ripened cheese. I. Isolation and assay. J. Food Sci. **26**, 72 (1961).

— — — Antimicrobial agent of aged surface-ripened cheese. II. Sources and properties of the active principle(s). J. Food Sci. **27**, 335 (1962).

—, R. O. WAGENAAR, and G. M. DACK: Inhibition of *Clostridium botulinum* and molds in aged, surface ripened cheese. Applied Microbiol. **7**, 33 (1959 a).

— — — Inhibition of *Clostridium botulinum* by culture filtrates of *Brevibacterium linens.* J. Bacteriol. **78**, 506 (1959 b).

— — — Relation of fatty acids to the inhibition of *Clostridium botulinum* in aged surface ripened cheese. Applied Microbiol. **7**, 228 (1959 c).

GREGORY, M. E., K. M. HENRY, and S. K. KON: Nutritive properties of freshly prepared and stored evaporated milks manufactured by a normal commercial procedure or by reduced thermal processes in the presence of nisin. J. Dairy Research **31**, 113 (1964).

GREENBERG, R. A., and J. H. SILLIKER: Spoilage patterns in *Clostridium botulinum* inoculated foods treated with tylosin. Bacteriol. Proc. **1962**, 21 (1962 a).

— — The action of tyolsin on spore-forming bacteria. J. Food Sci. **27**, 64 (1962 b).

— — The effect of tylosin on coagulase positive staphylococci in food products. J. Food Sci. **27**, 60 (1962 c).

GRIMM, A. I.: Use of antiseptics and antibiotics during storage of carrots. Sbornik Nauch. Rabot, Leningrad Inst. Sovet. Forgovli im F. Engel'sa **1959**, 17 (1959).

GROSOWICS, N., D. KAPLAN, and S. SCHNEERSON: Production of an antibiotic substance by a lactobacillus. 4th Internat. Congr. Microbiol., Rept. Proc. **1947**, 137 (1949).

GÜRSEL, A.: The bactericidal power of yoghurt against *Mycobacterium tuberculosis.* Turk Ijiyen Tecrubi Biyol Dergisi **17**, 50 (1957).

HAAS, G. J.: The influence of antibiotics on some biological contaminants in the brewery. Wallerstein Lab. Communications **18**, 253 (1955).

HALASZ, A., and E. JUHASZ: Activity changes of nisin due to heat and x-ray irradiation effects. Elelm. Ipar **17**, 160 (1963).

HALBEISEN, T.: An antibiotic substance isolated from *Cochlearia armoracia* (horseradish). Arzneimittel-Forsch. **7**, 321 (1957).

HALL, C. B.: Control of the darkening of butts of prepackaged celery. Proc. Amer. Soc. Hort. Sci. **81**, 354 (1962).

HALL, R. H.: Nisin. Brit. Pat. No. 916, 351 (1963).

HALLECK, F. E., O. C. BALL, and E. F. STIER: Factors affecting quality of prepacked meat. IV. Microbiological studies. C. Effects of initial bacteria count, kind of meat, storage temperature, antioxidants, and antibiotics on the rate of bacterial growth in packaged meat. Food Technol. **12**, 654 (1958).

HANSEN, H. C., G. E. WIGGINS, and J. C. BOYD: Modern methods of mastitis treatment cause trouble in the manufacture of cultured dairy products. J. Milk Food Technol. **13**, 359 (1950).

HARGROVE, R. E., H. E. WALTER, J. P. MALKAMES, and K. MASKELL: The effect of penicillin and streptomycin on Swiss cheese starters. J. Dairy Sci. **32**, 401 (1950).

HARMON, L. G.: Antibiotics in milk. Texas J. Sci. **3**, 167 (1951).

HARPER, W. J.: Antibiotic — protein interaction on the acid coagulation of milk. Amer. Milk Review **22**, 32 (1960).

HARRINGTON, G., and J. H. TAYLOR: The effect of antibiotic dietary supplements on the carcass measurements and dressing percentage of bacon pigs. J. Agr. Sci. **46**, 173 (1955).

HASHIDA, W.: Application of several antibiotics to the food industry. II. Application of several antibiotics to cows' milk. J. Ferment. Technol. (Japan) **31**, 15 (1953).

HASHIDA, W., and T. ASAI: Application of several antibiotics to the food industry. III. Application of the preservation of cows' milk. J. Ferment. Technol. (Japan) **31**, 122 (1953).

HAWLEY, H. B.: Nisin in food technology. Food Manuf. **32**, 370 (1957 a).

— Nisin in food technology. Food Manuf. **32**, 430 (1957 b).

— Nisin preparations. U. S. Pat. No. 2,785,108 (1957 c).

— The development and use of nisin. J. Appl. Bacteriol. **18**, 388 (1955).

— The permissibility and acceptability of nisin as food additive, Milchwissensch. **13**, 253 (1958).

—, and R. H. HALL: Nisin-containing product. Brit. Pat. No. 784,184 (1957).

— — Nisin. Brit. Pat. No. 844,782 (1960 a).

— — Nisin. U. S. Pat. No. 2,935,503 (1960 b).

HEDGECOCK, L. W., and L. R. JONES: Antibacterial activity of diacetyl. Bacteriol. Proc. **1950**, 127 (1950).

HEINEMANN, B., C. R. STUMBO, and A. SCHURLOCK: Use of nisin in preparing beverage-quality sterile chocolate-flavored milk. J. Dairy Sci. **47**, 8 (1964).

HEJLASZ, Z., M. KOCOT, and Z. ZAWADZKI: Effect of supravital administration of penicillin to meat animals on the prolongation of the meat shelflife. Med. Weterynar. **13**, 657 (1957).

HENNEBERG, O. H., and R. KISSLING: Accelerated ripening of beef and sheep meat at higher temperature when pretreated with aureomycin. Wien Tierärztl. Monatsschr. **46**, 697 (1959).

HIBBS, R. A., and J. C. BOYD: Some observations on testing milk samples for antibiotics. J. Milk Food Technol. **20**, 109 (1957).

HIETALA, P. K., and A. I. VIRTANEN: Precursors of benzoxazolinone in rye plants. II. Precursor I, the glucoside. Acta Chim. Scand. **14**, 502 (1960).

HIGASHI, S., and T. KANNO: Storage of wet sweet-potato starch. I. Effect of chlortetracycline on the storage of wet sweet-potato starch. Hakko Kogaku Zasshi **39**, 487 (1961).

HILLIG, F., L. R. SHELTON, JR., J. H. LOUGHREY, S. BETHEA, and C. M. CAMPBELL: Chemical indexes of decomposition in haddock stored in natural ice and chlortetracycline (CTC) ice. J. Assoc. Official Agr. Chemists **45**, 951 (1962).

HIRSCH, A.: The assay of the antibiotic nisin. J. Gen. Microbiol. **4**, 70 (1950).

— Growth and nisin production of a strain of *Streptococcus lactis*. J. Gen. Microbiol. **5**, 208 (1951).

—, and E. GRINSTED: The differentiation of the lactic streptococci and their antibiotics. J. Dairy Research **18**, 198 (1951).

— — Methods for the growth and enumeration of anaerobic spore-formers of cheese, with observations on the effect of nisin. J. Dairy Research **21**, 101 (1954).

—, and A. T. R. MATTICK: Some recent applications of nisin. Lancet **257**, 190 (1949).

—, and D. M. WHEATER: The production of antibiotics by streptococci. J. Dairy Research **18**, 193 (1951).

—, M. McCLINTOCK, and G. MOCQUOT: Observations on the influence of inhibitory substances produced by the lactobacilli of Gruyère cheese on the development of anaerobic spore-formers. J. Dairy Research **19**, 179 (1952).

—, E. GRINSTED, H. R. CHAPMAN, and A. T. R. MATTICK: A note on the inhibition of an anaerobic sporeformer in Swiss-type cheese by a nisin-producing streptococcus. J. Dairy Research **18**, 205 (1951).

HONKANEN, E., and A. I. VIRTANEN: Synthesis of some 1,4-benzoxazine derivatives and their antimicrobial activity. Acta Chim. Scand. **14**, 1214 (1960 a).

— — The synthesis of precursor II of benzoxazolinone formed in rye plants, and the enzymic hydrolysis of precursor I, the glucoside. Acta Chim. Scand. **14**, 504 (1960 b).

— — Unsaturated fatty acids in rye plants. Suomen Kemistilehti **B**, 33 (1960 c).

— — On the reaction of 2,4-dihydroxy-1,4-benzoxazin-3-one to 2,3-benzoxazolinone. Acta Chim. Scand. **15**, 221 (1961).

HORIE, S., K. SAHEKI, and Y. SEKINE: Studies on the putrefactive bacteria grown in the chlortetracycline (CTC) treated fish. Bull. Japan. Soc. Sci. Fisheries **27**, 188 (1961).
HOYLE, M., and A. A. NICHOLS: Inhibitory strains of lactic streptococci and their significance in the selection of cultures for starter. J. Dairy Research **15**, 398 (1948).
HUGLI, E., and I. PRUDENT: Effect of subtilin and streptomycin on the formation of nitric oxide hemochromogen in cooked ground fresh beef. Food Technol. **6**, 129 (1952).
HUMFELD, H.: Antibiotic activity of the fatty acid-like constituents of wheat bran. J. Bacteriol. **54**, 513 (1947).
HUNTER, G. J. E.: The effect of penicillin in milk on the manufacture of Cheddar cheese. J. Dairy Research **16**, 235 (1949 a).
— The effect of penicillin on lactic streptococci. J. Dairy Research **16**, 39 (1949 b).
IDLER, D. R., R. A. MACLEOD, and W. A. THOMPSON: Studies on the conversion of fish stickwater to solubles. II. Prevention of bacterial decomposition of stickwater at high temperatures. Applied Microbiol **3**, 205 (1955).
INOMOTO, Y., and W. HASHIDA: Application of several antibiotics to the food industry. I. Preservation of cows' milk. J. Ferment. Tech. **30**, 287 (1952).
IVANOVICS, G., and S. HORVATH: Isolation and properties of raphanin, an antibacterial substance from radish seed. Proc. Soc. Expt. Biol. Med. **66**, 625 (1947).
JACOBS, S. E., and A. H. DADD: Antibacterial substances in seed coats and their role in the infection of sweet peas by *Corynebacterium fascians* (Tilford) Dowson. Ann. Applied Biol. **47**, 666 (1959).
JACOBSON, M., J. SPENCER, M. HARD, and M. WELLER: The effects of antibiotic treatments, packaging films, and storage upon flavor, odor, and appearance of cut-up fryers. Poultry Sci. **38**, 40 (1959).
JACQUET, J., J. DELACROIX, G. LETOURNEUR, and O. VILLETTE: The sensitivity of lactic acid bacteria to penicillin in the manufacture of soft cheese. C. R. Acad. Agr. France **45**, 849 (1959).
JANKE, A., R. G. JANKE, K. BAUER, H. HEILMANN, H. KIRCHMEYER, and W. G. KOSTJAK: Increase in stability of the characteristics of chickens by treatment with chlortetracycline. Zentr. Bakteriol. Parasitenk., Abt. 2. **111**, 291 (1958).
JAY, J. M.: Effect of chlortetracycline on the microflora which develops during the spoilage of beef, and studies on the mode of action of this antibiotic in meat preservation. Dissertation Abstr. **17**, 949 (1957).
— Studies on the development of chlortetracycline-resistant staphylococci in chlortetracycline-treated beef. Antibiotics and Chemotherapy **10**, 441 (1960).
—, H. H. WEISER, and F. E. DEATHERAGE: Further studies on the preservation of beef with chlortetracycline. Food Technol. **11**, 563 (1957 a).
— — — Studies on the mode of action of chlortetracycline on the preservation of beef. Applied Microbiol. **5**, 400 (1957 b).
— — — The inhibition in beef of chlortetracycline-resistant bacteria by subbacteriostatic concentrations of the antibiotic. Applied Microbiol. **6**, 343 (1958).
JENSEN, L. B.: Antibiotic from avocado trees as a food preservative. U.S. Pat. No. 2,550,255 (1951 a).
— Antibiotic from broadleaf gum plant as a food preservative. U.S. Pat. No. 2,550,258 (1951 b).
— Antibiotic from Osage orange tree as a food preservative. U.S. Pat. No. 2,550,261 (1951 c).
— Antibiotic from pareira as a food preservative. U.S. Pat. No. 2,550,295 (1951 d).
— Antibiotic from rhatany as a food preservative. U.S. Pat. No. 2,550,256 (1951 e).
— Extraction of antibiotic from avocado tree. U.S. Pat. No. 2,550,254 (1951 f).
—, and W. R. HESS: Meat-curing pickle solution containing antibiotics. U.S. Pat. No. 2,550, 253 (1951).
—, and W. A. MILLER: Antibiotic from bittersweet as food preservative. U.S. Pat. No. 2,550,260 (1951 a).

JENSEN, L. B., and W. A. MILLER: Antibiotic from common agrimony as food preservative. U.S. Pat. No. 2,550,262 (1951 b).

— — Antibiotic from grapevine as a food preservative. U.S. Pat. No. 2,550,257 (1951 c).

— — Antibiotic from Iceland moss as a food preservative. U.S. Pat. No. 2,550,267 (1951 d).

— — Antibiotic from purple prairie clover as a food preservative. U.S. Pat. No. 2,550,264 (1951 e).

— — Antibiotic from St. John's wort as a food preservative. U.S. Pat. No. 2,550,266 (1951 f).

— — Antibiotic from spring avens as a food preservative. U.S. Pat. No. 2,550,265 (1951 g).

— — Antibiotic from white cedar as a food preservative. U.S. Pat. No. 2,550,263 (1951 h).

—, and J. E. SHERMAN: Antibiotic from the butternut tree as a food preservative. U.S. Pat. No. 2,550,268 (1951 a).

— — Antibiotic from galingale root as food preservative. U.S. Pat. No. 2,550,269 (1951 b).

JEZESKI, J. J., H. A. MORRIS, E. A. ZOTTOLA, E. GEORGE, and F. F. BUSTA: The effect of penicillin in milk on the growth and survival of *Staphylococcus aureus* during the manufacture and curing of Cheddar and Colby cheese. J. Dairy Sci. **44**, 1160 (1961).

JIMENIZ, A. V., R. P. FERNANDEZ, and M. E. GIRAUTA: Utilization of antibiotic and chelating substances in the treatment and storage of poultry (under refrigeration). Symp. Substances Etranger dans les Aliments, 6e, Madrid **1960**, 530 (1961).

JOHNS, C. K.: Differences in sensitivity of lactic starters to antibiotics. J. Dairy Sci. **36**, 1241 (1953).

— The action of penicillin on lactic acid streptococci in milk. J. Dairy Sci. **39**, 913 (1956).

—, and J. G. DESMARIS: The effect of residual penicillin in milk on the dye reduction tests for quality. Can. J. Agr. Sci. **33**, 91 (1953).

JOHNSTONE, D. B.: Effect of aureomycin upon acid production by cheese starters. Bacteriol. Proc. **1950**, 26 (1950).

KÄSTLI, P.: Defects of milk products through mastitis treated with penicillin. Schweiz. Arch. Tierheilk. **90**, 685 (1948).

KATAYAMA, T., and I. NAGAI: Chemical significance of the volatile components of spices from the food perservative viewpoint. III. Antibacterial activity of volatile components of nutmeg. J. Fac. Fisheries Animal Husbandry, Hiroshima Univ. **2**, 355 (1959).

— — Chemical significance of the volatile components of spices from the food preservation viewpoint. IV. Structure and antibacterial activity of some terpenes. Nippon Suisan Gakkaishi **26**, 29 (1960).

KATO, S., N. NISHIKAWA, and H. MUNEKATA: Microbiological studies of beer. I. Characters of microorganisms and effects of antibiotics in draught beer. Jôzô Kaguku Kenkyu Hokoku **3**, 52 (1957).

KATRANDZHIEV, K.: Inhibitory effect of *Lactobacillus bulgaricus* and *Streptococcus thermophilus* on *Brucella bovis* and *Erysipelothrix rhusiopathiae.* Nauchni Trudove **1**, 46 (1959).

— Study of the effects of *L. bulgaricus* and *Str. thermophilus* on coliform infections. Izv. Tsentr. Nauchnoizsled. Inst. po Zhivotnovudstvo. Sofiya **2**, 135 (1962).

KATZ, S. E., and K. HELRICH: The determination of penicillins and chlortetracycline in premixes and mixed feeds. Residue Reviews **7**, 74 (1964).

KATZNELSON, H., ana E. G. HOOD: Influence of penicillin and other antibiotics on lactic streptococci in starter cultures used in Cheddar cheese making. J. Dairy Sci. **32**, 961 (1949).

KAWABATA, T., G. SAKAGUCHI, and T. AKANO: Factors affecting the decomposition of oxytetracycline and chlortetracycline. J. Antibiotics Series **A 13**, 180 (1960).

KAYSER, K.: Is it advisable to alter methods hitherto prescribed and adopted in human milk banks? Zbl. Gynak 74, 455 (1952).
KAZAKOV, A. M., and V. K. DYKLOP: Antibiotics for improving the stability of meat and meat products. Tr. Vses. Nauchn. Issled. Inst. Myasn. Prom. **1955**, 30 (1955).
KELCH, F.: Aureomycin added to rations affecting quality of ham. Fleischwirtschaft **12**, 439 (1960).
KENNEDY, H. E.: Acid production in the presence of antibiotics. II. Effect of nutritional additives on oxytetracycline-or penicillin resistant cultures. J. Dairy Sci. **43**, 853 (1960).
KERSEY, R. C., F. C. VISOR, and C. L. WRENSHALL: Residual antibiotic levels in food products during storage and processing. Antibiotic Ann. **1953—1954**, 438 (1954).
KESSLER, B.: The ability of higher plants to synthesize antimicrobial substances. Arch. Biochem. Biophys. **55**, 287 (1955).
KLINE, E. A., J. KASTELIC, and G. C. ASHTON: The effect of vitamin B_{12}, cobalt, and antibiotic feeding on the composition of pork tissue of 100-pound pigs. J. Nutrit. **56**, 321 (1955).
KLINE, E. F., A. ABBEY, M. C. FIRINAN, and P. F. HOPPER: Thermal and non-thermal degradation of acronize chlortetracycline in fish and some shellfish. Food Technol. **14**, 305 (1960).
KLOSA, J.: Vegetable substances inhibiting seed germination and bacterial growth. Pharmazie **6**, 63 (1951).
KODAMA, R.: Studies on lactic acid bacteria. II. Lactolin, a new antibiotic substance, produced by lactic acid bacteria. J. Antibiotics **5**, 72 (1952).
KOHLER, A. R., H. P. BROQUIST, and W. H. MILLER: Chlortetraycline and the control of poultry spoilage. Food Technol. **8**, 19 (1954).
—, W. H. MILLER, and H. M. WINDLAN: Preservation and improved color retention of meats. U.S. Pat. No. 3,050,401 (1962).
KOOY, J. S.: Strains of *Lactobacillus plantarum* which inhibit the activity of the antibiotics produced by *Streptococcus lactis*. Netherlands Milk Dairy J. **6**, 323 (1952).
—, and J. W. PETTE: Inhibition of the growth of lactate-fermenting butyric acid bacteria by antibiotics from lactic acid streptococci. Netherlands Milk Dairy J. **6**, 302 (1952 a).
— — The inhibition of butyric acid fermentation in cheese by using antibiotic producing streptococci as starter. Netherlands Milk Dairy J. **6**, 317 (1952 b).
KOROLEVA, N.: Preservation of antibiotic activity of acidophilus bacteria under laboratory conditions. Molochn. Prom. **19**, 35 (1958).
KOSCIK, A.: The antibiotic properties of vegetable oils. Rocz. Akad. Med. Bialymstoku. **1**, 227 (1955).
KOVAL' CHUK, G. K., N. F. MOSKALENKO, and V. V. LIMANSKII: Use of chlortetracycline (biomycin) for preserving fresh goby, horse mackerel and anchovies. Tr. Azovo-Chernomorsk. Nauchn-Issled. Inst. Morsk. Rybn. Khoz. i Okeanogr. **1961**, 98 (1961).
KRAFT, A. A., L. E. ELLIOT, and A. W. BRYANT: Effect of antibiotic treatment on storage life of turkeys. Food. Technol. **12**, 660 (1958).
KRIENKE, W. A.: Drugs, mastitis and acid starters. Amer. Milk Rev. **12**, 54 (1950 a).
— Effects of various drugs in milk from mastitis-treated cows. S. Dairy Prod. J. **47**, 132 (1950 b).
— Penicillin in milk. Amer. Milk Review **11**, 24 (1950 c).
LABAW, G. D., and N. W. DESROSIER: Antibacterial activity of edible plant extracts. Food Research **18**, 186 (1953).
LAURENZA, A., and G. PIANESE: Effetti degli antibiotci sulla flora batterica del latte. Giorn. Bacteriol. Immunol. **49**, 297 (1956).
LAVIE, P.: Antibacterial substances in the honeybee colony. Ann. Abaille **3**, 103 and 201 (1960).
— Antibiotic substances in *Apis mellifica* and other social insects. Compt. Rend. **244**, 2653 (1957).

LAVIE, P.: Identification of antibacterial substances present in honey. Compt. Rend. **256**, 1858 (1963).
LEBLANC, F. R., K. A. DEVLIN, and C. R. STUMBO: Antibiotics in food preservation. I. The influence of subtilin on the thermal resistance of spores of *Clostridium botulinum* and the putrefactive anaerobe 3679. Food Technol. **7**, 181 (1953).
LEDFORD, R. A., and F. V. KOSIKOWSKI: Blue mold cheese as a potential agent of chronic penicillin urticaria. J. Dairy Sci. **44**, 1168 (1961).
LEJNIEKS, V., and H. G. SCHWIEGER: Antibiotics from garlic. Ger. Pat. No. 943,250 (1956).
LERKE, P. A., and L. FARBER: Prolongation of the keeping quality of fish and shellfish by antibiotics. Antibiotic Ann. **1956—1957**, 966 (1957).
— —, and W. HUBER: Preservation of fish and shellfish by relatively low doses of beta radiation and antibiotics. Food Technol. **15**, 145 (1961).
LEY, F. J., and R. H. HALL: Effect of gamma-radiation on nisin. Nature **188**, 1184 (1960).
LILLEY, M. D.: Food preservation with antibiotic nisin. Research Applied Ind. **11**, 472 (1958).
LIND, C.: The inhibition of butyric acid fermentation by nisin. Maelkeritidende **71**, 467 (1958).
LINNEWEH, F.: On the preservation of human milk. Med. Klin. **44**, 166 (1949 a).
— Streptomycin for conservation of human milk: clinical experiences. Med. Klin. **44**, 666 (1949 b).
LIPINSKA, E.: Microbiological methods of control of butyric acid fermentation in cheese. Acta Microbiol. Polon. **5**, 271 (1956 a).
— Prevention of the defects of cheese caused by butyric acid bacteria. Prace Inst. Przemyslu Mleczarskiego **2**, 1 (1956 b).
— Sugar fermentation in Edam cheese made with nisin — producing and nisin-resistant starters. Prace Inst. Przemyslu Mleczarskiego **8**, 1 (1961).
—, and M. STRZALKOWSKA: Preliminary studies on the use of antibiotics for the prevention of butyric acid fermentation in processed cheese. Prace Inst. Przemyslu Mleczarskiego **5**, 31 (1958).
— — Acclimatization of certain lactic starters to antibiotic products of *Streptococcus lactis*. 15th Internat. Dairy Congress, London **2**, 572 (1959).
— —, W. GOETTLICH, and W. SOLTYS: Application of nisin-resistant starters to the production of Edam cheeses. 16th Internat. Dairy Congress, Copenhagen **B**, 849 (1962).
LISKA, B. J.: A direct microscopic method for detecting antibiotic activity in milk. J. Dairy Sci. **42**, 905 (1959 a).
— Effects of penicillin on the morphology of *Streptococcus lactis, Streptococcus thermophilus*, and *Leuconostoc dextranicum*. J. Dairy Sci. **42**, 1391 (1959 b).
— A direct microscopic method for detecting antibiotic activity in milk. J. Milk Food Technol. **23**, 117 (1960).
LITTLE, J. E., and K. I. GRUBAUGH: Antibiotic activity of some crude plant juices. J. Bacteriol. **52**, 587 (1946).
LIZGUNOVA, A. V.: Antibacterial properties of gramicidin in milk. Gigiena i Sanit. **1952**, 27 (1952).
LOCKHART, W. C.: Eggs not contaminated by drugs in feed. N. D. Agr. Expt. Farm Research Bull. **22**, 11 (1962).
LORCH, L. V.: Method for obtaining microbial suspensions with heterologous resistance to antibacterials. Rend. Ist. Super. Sanita. **19**, 598 (1956).
—, R. NEGRI, and G. PENSO: Occurrence of fermentations in the presence of microorganisms with induced antibiotic resistance and under antibiotic action. Sci. Rept. 1st Super. Sanita. **1**, 397 (1961).
— — —, and P. SAVI: Process to treat food products by the use in presence of antibiotics and either antibiotic resistant or antibiotic dependent bacteria strains. U.S. Pat. No. 3,098,744 (1963).
LOVELL, T. H.: Bactericidal effects of onion vapors. Food Research **2**, 435 (1937).
LUDWIG, R. A., E. Y. SPENCER, and C. H. UNIVIN: An antifungal factor from barley of possible significance in disease resistance. Can. J. Bot. **38**, 21 (1960).

LUKOVNIKOVA, L.: Antibiotic properties of lactic streptococci. Molochn. Prom. **22,** 39 (1961).

MALIN, B., R. DOBBS, and L. MARTIN: The effect of tylosin lactate on *Clostridium botulinum* type E. Proc. 24th Ann. Meeting Inst. Food Technol. **1964,** 93 (1964).

MARTH, E. H.: Antibiotics in milk — a review. I. Recent developments. J. Milk Food Technol. **24,** 36 (1961 a).

— Antibiotics in milk — a review. II. Methods for detection of antibiotics in milk. J. Milk Food Technol. **24,** 70 (1961 b).

—, and B. E. ELLICKSON: Antibiotic residues in milk and milk products. A review. J. Milk Food Technol. **22,** 241 (1959 a).

— — Problems created by the presence of antibiotics in milk and milk products. A review. J. Milk Food Technol. **22,** 266 (1959 b).

—, and R. V. HUSSONG: Effect of skimmilks cultured with different strains of *Leuconostoc citrovorum* on growth of some microorganisms associated with cottage cheese spoilage. J. Dairy Sci. **45,** 652 (1962).

— — Effect of skimmilk cultured with different strains of *Leuconostoc citrovorum* on growth of some bacteria and yeasts. J. Dairy Sci. **46,** 1033 (1963).

—, F. J. ALEXANDER, and R. V. HUSSONG: Studies on disc assay methods for detection of antibiotics in milk. J. Milk Food Technol. **26,** 150 (1963).

MARTIN, J. H., and W. J. HARPER: Factors affecting the loss of antibiotic activity in milk. Antibiotic Ann. **1958—1959,** 890 (1959).

MARUZZELLA, J. C., and M. FREUNDLICH: Antimicrobial substances from seeds. J. Amer. Pharm. Assoc. **48,** 356 (1959).

MASILUNGAN, V. A., P. A. BERBA, N. C. DIOKNO, and F. B. BORILLO: The antimicrobial activity of extracts from acacia leaves. Philippine J. Sci. **87,** 7 (1958).

—, N. C. DIOKNO, and V. B. QUISUMBING: Screening of Philippine medicinal plants used in the treatment of tuberculosis for substances inhibitory to *Mycobacterium tuberculosis* 607. Philippine J. Sci. **88,** 245 (1959).

—, J. MARANON, V. V. VALENCIA, N. C. DIOKNO, and P. DELEON: Screening of Philippine higher plants for antibacterial substances. Philippine J. Sci. **84,** 275 (1955).

MATHER, D. W., and F. J. BABEL: Inhibition of certain types of bacterial spoilage in creamed cottage cheese by use of a creaming mixture prepared with *Streptococcus citrovorus.* J. Dairy Sci. **42,** 1917 (1959).

MATIKKLA, E. J., and A. I. VIRTANEN: A new sulfur-containing amino acid in onion. II. Suomen Kemistilehti B **30,** 219 (1957).

MATSUI, T., G. TOKUTOMI, A. TAKASE, Y. AKAO, R. SANO, T. BITO, M. NAKAJIMA, H. WATANABE, and T. FUKAZAWA: Putrefaction prevention experiment on meat by using an antibiotic (oxytetracycline). Bull. Inst. Public Health (Tokyo) **8,** 135 (1959).

MATTIK, A. T. R., and A. HIRSCH: A powerful inhibitory substance produced by group N. streptococci. Nature **154,** 551 (1944).

— — Further observations on an inhibitory substance (nisin) from lactic streptococci. Lancet **253,** 5 (1947).

— — Manufacture and preservation of cheese. U.S. Pat. No. 2,744,827 (1956).

— — Improvements in or relating to the manufacture or preservation of cheese. Brit. Pat. No. 713,251 (1954).

— — The streptococci and antibiotics. Proc. 12th Internat. Dairy Congress Stockholm **2,** 546 (1949).

MCCLINTOCK, M. L. SERRES, J. J. MARZOLF, A. HIRSCH, and G. MOCQUOT: The inhibitory action of nisin-producing streptococci on the development of anaerobic spore-bearers in processed Gruyère cheese. J. Dairy Research **19,** 187 (1952).

MCKEE, R. C., J. CONKEY, and J. A. CARLSON: A study of the comparative shelf life of wet and dry packed poultry. Poultry Sci. **38,** 260 (1959).

MCVICKER, R. J., L. E. DAWSON, W. L. MALLMANN, S. WALTER, and E. JONES: The effect of certain bacterial inhibitors on shelf-life of fresh fryers. Food Technol. **12,** 147 (1958).

— — — — — The effect of chlortetracycline on shelf-life of fresh poultry meat treated in a commercial processing plant. Poultry Sci. **38,** 337 (1959).

MEANWELL, L. J.: The influence of penicillin on cheese starter activity. J. Applied Bacteriol. 25, 128 (1962).
MEEWES, K. H., and S. MILOSEVIC: Culture filtrates of *Lactobacillus helveticus* and their influence on the gas-producing ability of *Escherichia coli*. Milchwissensch. 17, 678 (1962).
—, and S. PASLAWSKI: Effect of antibiotics on lactic acid bacteria. Kieler Milchwirtsch. Forschungsber. 3, 543 (1951).
MEREDITH, W. E., H. H. WEISER, and A. R. WINTER: Chlortetracycline and oxytetracycline residues in poultry tissues and eggs. Applied Microbiol. 13, 86 (1965).
MEYER, R. C., A. R. WINTER, and H. H. WEISER: Edible protective coatings for extending the shelf life of poultry. Food Technol. 13, 146 (1959).
MICHENER, H. D.: Effect of subtilin on ungerminated bacterial spores. Bacteriol. Proc. 1953, 29 (1953 a).
— The bacterial action of subtilin on *Bacillus stearothermophilus*. Applied Microbiol. 1, 215 (1953 b).
— The action of subtilin on heated bacterial spores. J. Bacteriol. 70, 192 (1955).
MIKOLAJCIK, E. M.: Physiology of oxytetracycline resistant and sensitive *Streptococcus lactis*. Dissertation Abstr. 20, 3471 (1960).
—, and I. A. GOULD: Host-bacteriophage relationship as influenced by penicillin. J. Dairy Sci. 45, 653 (1962).
—, W. J. HARPER, and H. E. KENNEDY: Acid production in the presence of antibiotics. I. Oxytetracycline inactivating properties of a resistant strain of *Streptococcus lactis*. J. Dairy Sci. 43, 853 (1960).
MILLER, W. A., and R. W. MORRISON: The microbiology of formerly clean and dirty, non-segregated, washed, early spring and early summer eggs. Poultry Sci. 37, 1022 (1958).
MILLER, W. H.: Antibiotic introduced as spoilage inhibitor for fresh poultry. Food Eng. 28, 43 and 194 (1956).
MOCQUOT, G., and E. LEFEBVRE: A simple procedure to detect nisin in cheese. J. Appl. Bacteriol. 19, 322 (1956).
MOELLER-MADSEN, A., and G. BENDTSEN: Nogle antibiotica's haemmenda virkning paa almindeslige syrevaekker, *Streptococcus thermophilus* og *Thermobacterium helvetium*. Danish Govt. Research Inst. on Dairy Industry. Bull. 105, p. 72 (1956).
MORR, C. V.: Effect of penicillin and aureomycin upon the phosphatase enzyme system of milk. J. Dairy Sci. 42, 898 (1959).
MOYCHO, W., and W. GROMSKA: Antagonistic action of *Streptococcus lactis* against *Bacillus subtilis* and *Pseudomonas fluorescens* observed in milk. Acta Microbiol. Polon. 5, 267 (1956).
MOYER, R. H., B. A. SOUTHCOTT, and H. L. A. TARR: Distribution of chlortetracycline (CTC) antibiotic in ice made from hard waters. Progress Rept. Pacific Coast Stations Fisheries Research Board of Canada. No. 112, pp. 21—22 (1958).
MULLER, W. H.: Influence of antibiotics on microorganisms causing fruit and vegetable rots. Amer. J. Botany 45, 183 (1958).
MYERS, R. P.: Antibiotic residues in milk. Residue Reviews 7, 9 (1964).
NAGEL, V.: Preservation of raw milk with streptomycin. Ärztl. Wochenschr. 5, 393 (1950).
NERI, A.: Bactericidal properties of some strains of lactobacilli. Latte 35, 518 (1961).
NEUWEILER, W., and P. KÄSTLI: Aureomycin treatment of puerperal mastitis. Gynaecologia 135, 1 (1953).
NEWTON, G. G. F., E. P. ABRAHAM, and N. J. BERRIDGE: Sulfur-containing amino acids of nisin. Nature 171, 606 (1953).
NG, H., R. H. VAUGHN, G. F. STEWART, C. W. NEGEL, and K. SIMPSON: Antibiotics in poultry meat preservation: Development of resistance among spoilage organisms. Applied. Microbiol. 5, 331 (1957).

NILSON, K. M., M. CARRANCEDO, and K. M. SHAHANI: The use of antifungal agents in the retardation of cottage cheese spoilage. J. Dairy Sci. **43,** 842 (1960).

NILSSON, G., and R. NILSSON: Inhibitory effect of Swedish ropy milk on fungi. Ann. Acad. Sci. Fennicae, Series A II. **60,** 235 (1955).

NIVEN, C. F., and W. R. CHESBRO: Antibiotics and irradiation in meat preservation. Proc. 8th Conf. on Research, Council on Research, Amer. Meat Inst., Univ. of Chicago **1956,** 47 (1956).

NJOKU-OBI, A. N., J. V. SPENCER, E. A. SAUTER, and M. W. EKLUND: A study of the fungal flora of spoiled chlortetracycline treated chicken meat. Applied Microbiol. **5,** 319 (1957).

NOVAK, A. F., E. A. FIEGER, and M. E. BAILEY: Chlortetracycline for preserving Gulf oysters. Food Technol. **12,** 237 (1958).

— —, and K. A. STOLZE: In vitro effects of chlortetracycline on bacteria indigenous to Gulf shrimp and oysters. Food Technol. **14,** 585 (1960).

NOVAK, L. J.: Preservation of frozen foods. U.S. Pat. No. 2,790,720 (1957).

O'BRIEN, R. T., and D. S. TITUS: The effect of subtilin on heat-activated and severely heated spores of a putrefactive anaerobic bacterium. J. Bacteriol **70,** 487 (1955).

— —, K. A. DEVLIN, C. R. STUMBO, and J. C. LEWIS: Antibiotics in food preservation. II. The influence of subtilin and nisin on the thermal resistance of food spoilage bacteria. Food Technol. **10,** 352 (1956).

OCKLITZ, H. W., and E. F. SCHMIDT: Preservation of human milk with aureomycin. Arch. Kinderheilk. **142,** 21 (1951).

OKAZAKI, K., and S. OSHIMA: Antibacterial activity of higher plants. XXV. Antimicrobial effect of essential oils. J. Pharm. Soc. Japan **73,** 690 (1953).

OKITSU, T.: Food preservatives. I. Influence of temperature and hydrogen-ion concentration on the action of food preservatives. Nippon Suisan Gakkaishi **26,** 797 (1960 a).

— Food preservatives. II. Effects of certain food preservatives and antibiotics on the growth and histamine production of *Proteus morganii.* Nippon Suisan Gakkaishi **26,** 843 (1960 b).

OLSON, J. C. Jr., D. S. WILLOGHBY, E. L. THOMAS, and H. A. MORRIS: Observations on the keeping quality of pasteurized milk with and without added aureomycin, which was obtained from various sources and stored at 45° F. J. Dairy Sci. **35,** 485 (1952).

— — — — The keeping quality of pasteurized milk as influenced by the growth of psychrophilic bacteria and the addition of aureomycin. J. Milk Food Technol. **16,** 213 (1953).

OSHIMA, T., and F. UJIIE: The use of antibiotics in the preservation of beef. Japanese J. Vet. Sci. **21,** 61 (1959).

OVERBY, A.: The effect of various antibiotics in milk following intramammary infusion. Nord. Veterinarmed. **4,** 993 (1952).

OXFORD, A. E.: Diplococcin, an antibacterial protein elaborated by certain milk streptococci. Biochem. J. **38,** 178 (1944).

PAGANO, J. F., A. E. TEMPLE, and H. YACOWITY: Retarding the spoilage of meat products. U.S. Pat. No. 2,944,907 (1960).

PEDERSON, C. S., and P. FISHER: The bactericidal action of cabbage and other vegetable juices. N. Y. State Agr. Expt. Station Tech. Bull. No. 273 (1944).

PETTE, J. W., and J. S. KOOY: The inhibition of butyric acid fermentation in cheese by antibiotic producing starter. 13th Internat. Dairy Congress, The Hague. **3,** 1172 (1953).

PEYNAUD, E., and S. LAFOURCADE: The study of a new antibiotic and antiseptic agent against *Saccharomyces.* Compt. Rend. **236,** 1924 (1953).

Pfizer, Chas. and Co., Inc.: Preservation of fresh meats and fish. Brit. Pat. No. 880,687 (1961).

PHILLIPS, A. W., H. R. NEWCOMB, and F. C. BACH: Experimental preservation of poultry by chlortetracycline and radiation. Food Technol. **14,** 521 (1960).

PHILLIPS, A. W., H. R. NEWCOMB, T. ROBINSON, F. C. BACH, W. L. CLARK, and A. R. WHITEHILL: Experimental preservation of fresh beef with antibiotics and radiation. Food Technol. **15**, 13 (1961).

PIVNYAK, I. G., and A. S. OBRAZTSOVA: Fungal mycelia in the raising and fattening of agricultural animals. Primenenie Antibiotikov v Zhivotnovodstve, Materialy Soveshch. Otd. Zhivotnovodstva, Vses. Akad. Sel'skokhoz. Nauk. Moscow **1958**, 60 (1959).

POETSCHKE, G.: Disinfection and preservation of human milk with streptomycin. Klin. Wochschr. **27**, 476 (1949).

— Disinfection and preservation of raw milk with streptomycin. Mschr. Kinderheilk. **98**, 177 (1950).

— Milk product containing an antibiotic. Brit. Pat. No. 688,968 (1953).

POLONSKAYA, M. S.: Antibiotic substances produced by *Lactobacillus acidophilus*. Mikrobiologiya **21**, 303 (1952).

—, V. V. LEONOVICH, and M. P. BIBERDIEVA: Concentrate of bacteriostatic substances produced by acidophilic bacteria. Dokl. Vses. Akad. Nauk. **1953**, No. 8, 21 (1953).

POOLE, G., and B. MALIN: Some aspects of the action of tylosin on *Clostridium* species PA 3679. J. Food Sci. **29**, 475 (1964).

POTHMANN, F. J.: The influence of natural honey on the growth of the tubercle bacillus. Z. Hyg. Infektionskrankh. **130**, 468 (1950).

PRICE, K. E., Z. ZOLLI, Jr., J. C. ATKINSON, and H. G. LUTHER: Antibiotic inhibitors. I. The effect of certain milk constituents. Antibiotics and Chemotherapy **7**, 672 (1957).

PRICHRA, M.: Über die bactericide Wirkung des Naturhonigs. Z. Hyg. Infektionskrankh. **120**, 437 (1938).

PROST, E.: Evaluation of the usefulness of antibiotics in the production of canned meat, esp. chopped pork. Ann. Univ. Mariae Curie-Skledowska Sect. DD-Vet. Sci. **8**, 49 (1955).

PRUTHI, J. S., L. GIRDHARI, and V. SUBRAHMANYAN: Chemistry and technology of garlic powder. Food Sci. (Mysore) **8**, 429 (1959 a).

—, L. J. SINGH, K. INDIRAMMA, A. N. SANKRAN, and G. LAL: Effect of nitrogen packing and storage temperature on the quality of garlic powder. Food Sci. (Mysore) **8**, 461 (1959 b).

— —, S. S. KALBAG, and G. LAL: Effect of different methods of dehydration on the quality of garlic powder. Food Sci. (Mysore) **8**, 444 (1959 c).

— —, and G. LAL: Determination of the critical temperature of dehydration of garlic. Food Sci. (Mysore) **8**, 436 (1959 d).

— — — Some technological aspects of dehydration of garlic; some factors affecting the quality of garlic powder during dehydration. Food Sci. (Mysore) **8**, 441 (1959 e).

PULAY, G.: Research on *Streptococcus lactis* strains producing antibiotics against clostridia. 14th Internat. Dairy Congress, Rome **2** (part 2), 432 (1956).

—, and R. Z. BITTERA: Prevention of mold growth on the surface of cheeses with flavofungin, a new antifungal antibiotic. 15th Internat. Dairy Congress, London **2**, 933 (1959).

RAIBAUD, P.: The use of a thermophilic bacterium for the rapid determination of nisin. Ann. Inst. Natl. Rèch. Agron., Series E. **5**, 441 (1956).

— Evidence for the fixation of nisin by the cells of a thermophilic bacteria used in the determination of the dose of that antibiotic. Ann. Inst. Natl. Rèch. Agron., Series E., Ann. Technol. Agr. **6**, 409 (1957).

RAMSEIER, H. R.: The action of coliform bacteria on the stability of nisin with special regard to the influence of the test medium. 15th Internat. Dairy Congress, London **2**, 566 (1959).

— The influence of nisin on *Clostridium butyricum*. Prazm. Arch. Mikrobiol. **37**, 57 (1960).

— Zur Anwendung von Nisin in der Milchwirtschaft. Wissenschaftl. Beilage Nr. 74, **1961**, 1 (1961).

RAO, P. L., and S. C. L. VERMA: Antibiotic principle of *Allium sativum*. Structure of allicin; preparation and properties of diphenyl disulfide oxide. J. Indian Inst. Sci. **34**, 315 (1952).
REHM, H. J.: Action of preservative combinations. IV. Action of simple combinations of preservatives with some antibiotics on *Escherichia coli*. Z. Lebensm. Untersuch. u. Forsch. **113**, 144 (1960).
RIBERAU-GAYON, J., and E. PEYNAUD: Inhibition of wine yeasts by vitamin K_5 and other antibiotics. Compt. Rend. Acad. Agr. France **39**, 479 (1952).
RICHARDS, R. J.: Sensitivity to penicillin of some cheese-starter cultures. Australian J. Dairy Technol. **13**, 122 (1958).
— The effect of antibiotics on lactic streptococci and lactic streptococcal host-phage relationships. Dissertation Abstr. **21**, 2859 (1961).
—, and H. E. KENNEDY: Antibiotic resistance of starter cultures belonging to *Streptococcus*. J. Dairy Sci. **43**, 852 (1960).
RITTER, P.: Slow acid production in hard cheese. Schweiz. Milchz. **82**, 655 (1956 a).
— The antagonism between lactic acid bacteria and bacteria of the coliaerogenes group. II. The development of specific antibiotic substances by lactic acid bacteria. Schweiz. Milchz. **82** (Sci. Suppl. No. 35), 273 (1956 b).
— The use of nisin or nisin-forming cultures in the manufacture of Emmental cheese. Schweiz. Z. allgem. Pathol. u. Bakteriol. **20**, 548 (1957).
— Antibiotic action of enterobacteria on lactic acid bacteria in cheese. 15th Internat. Dairy Congress, London **2**, 587 (1959).
— Nisin synthesis by *Streptococcus lactis*. Pathol. Microbiol. Basle **26**, 564 (1963).
RITTER, W.: Über das Verhalten einiger Organismen aus Rahmsäuerungskulturen. Schweiz. Milchz. **71**, 255 (1945).
ROBINSON, W. L., L. E. KUNKLE, and V. R. CAHILL: The use of an antibiotic in rations for hogs. Ohio Agr. Expt. Station, Wooster, Ohio. Research Bull. 769 (1956).
ROGERS, L. A.: The inhibitory effect of *Streptococcus lactis* on *Lactobacillus bulgaricus*. J. Bacteriol. **16**, 321 (1928).
ROOS, H.: The preservation of raw human milk. Mschr. Kinderheilk. **99**, 66 (1951).
—, and M. KINDLER: Preservation of human milk with citric acid or streptomycin. Mschr. Kinderheilk. **97**, 494 (1950).
ROSELL, J., and S. MATALLANA: Experimental results on the use of nisin-producing cultures for the prevention of butyric fermentation in cheese. Sci. Aliment **2**, 126 (1956).
ROYCE, A., C. BOWLER, and G. SYKES: The assay of penicillin using penicillinase. J. Pharm. Pharmacol. **4**, 904 (1952).
ROZANSKY, R., and A. BRZEZINSKY: The excretion of penicillin in human milk. J. Lab. Clinical Med. **34**, 497 (1949).
ROZMANOVA, N. V.: Effect of antibiotics on the bacterial flora in yeast production. Izv. Vysshikh Zavedenii. Pishchivaya Tecknol. **6**, 41 (1958).
RUSOFF, L. L., G. C. LEE, and E. J. STONE: Aureomycin (chlortetracycline) distribution in milk. J. Dairy Sci. **40**, 1390 (1957).
SABINE, D. B.: An antibiotic-like effect of *Lactobacillus acidophilus*. Nature **199**, 811 (1963).
SAKAGUCHI, G., S. SAKAGUCHI, T. KAWABATA, Y. NAKAMURA, T. AKANO, and K. SHIROMIZU: Influence of oxytetracycline upon the toxin production of type E *Clostridium botulinum*. Japan J. Med. Sci. Biol. **13**, 13 (1960).
SALUNKHE, D. K., G. M. COOPER, A. S. DHALIWAL, A. A. BOE, and R. L. RIVERS: On storage of fruits; effects of pre- and post-harvest treatments. Food Technol. **16**, 119 (1962).
SARTORY, R., J. MEYER, and J. MALGRAS: Propionic fermentation and acid-fast bacteria in hard cheese. Bull. Acad. Nat. Med. Paris **135**, 708 (1951).
SASAYAMA, S., S. HAMADA, K. YAMADA, Y. YONE, and T. TOMIYAMA: Change in freshness and chlortetracycline residues in chlortetracycline-treated trawler's catches due to different species, size and method of treatment. Nippon Suisan Gakkaishi **26**, 361 (1960).
SATO, M.: Effects of antibiotics on contamination for sake — moromi mash. Nippon Nogei Kagaku Kaishi **32**, 119 (1958).

SAVICH, A. L., and C. E. JANSEN: Improving the color of meat. U.S. Pat. No. 2,798,812 (1957).
SCHEPARTZ, A. I., and M. H. SUBERS: The glucose oxidase of honey. I. Purification and some general properties of the enzyme. Biochem. Biophys. Acta **85**, 228 (1964).
SCHIESSLER, P.: Die Wirkung des Chloromycetins bei Eutererkrankungen. Milchwissenschaft. Ber. **2**, 41 (1952).
SCHMIDT, B., and G. HANNEMANN: The in vitro inhibition of *Salmonella* and *Escherichia* spp. by fermented milks. Z. Hyg. Infektionskrankh. **145**, 499 (1959).
—, H. MODDE, and W. DINTER: Inhibition of the antibiotic activity of streptomycin and dihydrostreptomycin by human milk. Arch. Hyg. Bakteriol. **144**, 433 (1960).
SCHMIDT, F. J., and W. J. STADELMAN: Effects of antibiotics and heat-treatment of shell eggs on quality after storage. Poultry Sci. **36**, 1023 (1957).
SCOTT, R.: Destruction of starter activity by penicillin. Dairy Ind. **15**, 1260 (1950).
SCOTT, W. E., H. McKAY, P. S. SHAFFER, and T. D. FONTAINE: The partial purification and properties of antibiotic substances from the banana *(Musa sapientum)* J. Clinical Invest. **28**, 899 (1949).
SEBASTIO, C., and S. CARACCIOLO: Antibiotic use in fish preservation. IV. Chemical research on Acronize-treated sardines. Atti. Soc. Ital. Sci. Vet. **11**, 612 (1957).
—, and V. ORLANDI: Observations on the action of tetracyclines in the preservation of Adriatic codfish. IV. Chemical characteristics. Atti. Soc. Ital. Sci. Vet. **12**, 467 (1958).
SEHGAL, J. M.: Antimicrobial substances from flowering plants. Hindustan Antibiotic Bull. **4**, 3 (1961).
SEITZ, E. W., J. A. ELLIOTT, and C. K. JOHNS: Inhibition of staphylococci by the antibiotic Tylosin in Cheddar cheese making. Bact. Proc. **1963**, 16 (1963).
SENN, D. P., K. VISMESWARIAH, and N. L. LAHIRY: Studies on dry-salting and sundrying of mackerel *(Rastrelliger kanagurtha)*. II. Effect of chlortetracycline, sorbic acid, sodium propionate, sodium benzoate, and sodium acid phosphate on the keeping quality of sun-dried mackerel. Food Sci. (Mysore) **10**, 132 (1960).
SENECA, H., E. HENDERSON, and A. COLLINS: Bactericidal properties of yoghurt. Amer. Practitioner Dig. Treat. **1**, 1252 (1950).
SHANHANI, K. M.: Factors affecting terramycin activity in milk, broth, buffer and water. J. Dairy Sci. **41**, 382 (1958).
— Nucleic acid and nucleoprotein metabolism of terramycin-sensitive and resistant organisms. J. Dairy Sci. **40**, 603 (1957 a).
— The effect of heat and storage on the stability of aureomycin in milk, buffer, and water. J. Dairy Sci. **40**, 289 (1957 b).
— Comparison of nucleic acids and other phosphorus compounds of terramycin-sensitive and -resistant strains of *Streptococcus lactis*. J. Dairy Sci. **42**, 1659 (1959).
— Galactokinase and galactowaldenase enzyme activity of *Streptococcus lactis*. J. Dairy Sci. **43**, 852 (1960).
— Inhibitory effect of nisin on various organisms. J. Dairy Sci. **45**, 827 (1962).
—, and R. C. CHANDAN: Sensitivity of milk lipase to antibiotics. J. Dairy Sci. **45**, 1178 (1962).
—, I. A. GOULD, H. H. WEISER, and W. L. SLATTER: Effect of heat treatments on antibiotic content of milk. J. Dairy Sci. **37**, 647 (1954).
— — — — Observations on antibiotics in a market milk supply and the effect of certain antibiotics on the keeping quality of milk. Antibiotics and Chemotherapy **6**, 544 (1956).
—, and W. J. HARPER: Differences in the phosphorus metabolism of terramycin sensitive and resistant organisms. J. Dairy Sci. **39**, 915 (1956).
— — Differences in the acid production and metabolism of oxytetracycline sensitive and resistant organisms. Applied Microbiol. **6**, 9 (1958 a).

SHANHANI, K. M., and W. J. HARPER: The development of antibiotic resistance in cheese starter cultures. Milk Prod. J. **49,** 15 and 53 (1958 b).
—, K. M. NILSON, and P. A. DOWNS: Effects of antifungal antibiotics upon the keeping quality of cottage cheese. 15th Internat. Dairy Congress, London **2,** 926 (1959).
—, and J. R. VAKIL: Carbohydrate metabolism of lactic acid cultures. I. Effect of antibiotics on the glucose, galactose and lactose metabolism of *Streptococcus lactis.* J. Dairy Sci. **45,** 1057 (1962 a).
— — Certain enzymes of glycolytic and hexosemonophosphate shunt pathways of *Streptococcus lactis.* J. Dairy Sci. **45,** 655 (1962 b).
SHANNON, W. G., and W. J. STADELMAN: The efficacy of chlortetracycline at several temperatures in controlling spoilage of poultry meat. Poultry Sci. **36,** 121 (1957).
SHERMAN, J. M., and H. M. HODGE: The bactericidal properties of certain plant juices. J. Bacteriol. **31,** 96 (1936).
SHIRK, R. J., and W. L. CLARK: The effect of pimaricin in retarding the spoilage of fresh orange juice. Food Technol. **17,** 108 (1963).
—, E. F. KLINE, and W. L. CLARK: The effect of pimaricin in the control of yeast growth in orange juice concentrates. Proc. 24th Ann. Meeting Inst. Food Technol. **1964,** 104 (1964).
—, A. R. WHITEHILL, and W. L. CLARK: Bioassay of pimaricin and its binding effect in orange juice. J. Food. Sci. **27,** 605 (1962).
—, A. R. WHITEHILL, and L. R. HINES: Degradation product in cooked chlortetracycline treated poultry. Antibiot. Ann. **1956—1957,** 843 (1957).
SHRIMPTON, D. H., E. M. BARNES, and W. S. MILLER: Control of spoilage of eviscerated poultry carcasses by treatment with antibiotics before slaughter. J. Sci. Food Agr. **9,** 353 (1958).
SILVA, M., and H. R. BUCKLEY: Activity of egg white against fungi pathogenic to man. In: Fungi and fungous diseases, G. Dalldoz, ed. P. 272. Springfield, Ill.: Charles C. Thomas 1962.
SILVESTRINI, D. A., G. W. ANDERSON, and E. S. SNYDER: Chlortetracycline as related to the microbiology and preservation of poultry meat. Poultry Sci. **37,** 179 (1958).
— — — Effect of tetracycline on the preservation of poultry processed by two commercial methods. Poultry Sci. **38,** 132 (1959).
SJÖSTRÖM, G., and S. WILLART: Effect of nisin on butyric acid bacteria of lactose and lactate fermenting types. Svenska Mejeritidn. **45,** 237 and 247 (1953).
SKORODUMOVA, A. M.: Antibiotic properties of fermented milk beverages. Voprosy Pitaniya **15,** 32 (1956).
SLEETH, R. B.: Effect of environmental conditions during aging on shrinkage and organoleptic characteristics of beef and the efficacy of oxytetracycline for aging beef. Dissertation Abstr. **20,** 3686 (1960).
—, J. C. ARMSTRONG, H. S. GOLDBERG, and H. D. NAUMANN: Antibiotic preservation of beef with subsequent feeding to experimental animals. Food Technol. **14,** 505 (1960).
SMITH, K. L.: Penicillin in milk used for cottage cheese making. Dissertation Abstr. **21,** 2437 (1961).
SMITH, W. L. Jr.: Streptomycin sulfate for control of bacterial soft rot of packaged spinach. Phytopathol. **42,** 475 (1952).
SOUTHCOTT, B. A., and H. L. A. TARR: Magnesium as a factor in determining the comparative sensitivity of marine bacteria to four tetracycline antibiotics. Can. J. Microbiol. **7,** 284 (1961).
—, E. G. BAKER, J. W. BOYD, and H. L. A. TARR: Comparative effectiveness of tetracycline antibiotics for fish preservation. Food Technol. **12,** 108 (1958 a).
—, R. MOYER, E. G. BAKER, and H. L. A. TARR: Experiments on preservation of fish with tetracycline antibiotics. Progress Rept. Fisheries Research Board of Canada Pacific Coast Station, No. **110,** 16 (1958 b).
SPARE, C. G., and A. I. VIRTANEN: The volatile carbonyls and alcohols in the flavour substances of onion *(Allium cepa).* Acta Chim. Scand. **15,** 1280 (1961).

STADELMAN, W. J., W. W. MARION, and M. L. ELLER: Antibiotic preservation of fresh poultry meat. Antibiot. Ann. **1956—1957**, 839 (1957).
STEEL, R., and T. K. WALKER: Studies on the antibacterial activity of certain strains of *Acetobacter*. J. Gen. Microbiol. **18**, 369 (1958).
STEINER, G., and H. L. A. TARR: Penetration of chlortetracycline into fish muscle and its destruction by heat. Can. J. Technol. **34**, 215 (1956).
STERN, J. A., H. L. LIEBMAN, G. KUDO, J. G. CHAPEL, R. A. OLSEN, L. L. FARBER, and M. GRENNAN: Potential application of antibiotics in the salmon canning industry. II. Chemical and bacteriological evaluations. Food Technol. **12**, 132 (1958).
STOCKER, W.: Yeast as a cause of blowing in cheese-lactic acid bacteria as inhibitors. Deut. Molkerei-Ztg. **75**, 131 (1954).
STOLTZ, E. I., and D. J. HANKINSON: Antibiotics and lactic acid starter cultures. Applied Microbiol. **1**, 24 (1953 a).
— — The effects of antibiotics on the bacterial plate counts of normal raw milk. J. Milk Food Technol. **16**, 157 (1953 b).
— — The effects of antibiotics in milk on the human intestinal coliform bacteria. J. Milk Food Technol. **17**, 76 (1954).
STORGARDS, T.: Nachweis von Penicillin und anderen Antibiotika in Milch. Milchwissensch. **17**, 369 (1962).
STORRS, F. C., and W. HIETT BROWN: The incidence of penicillin in milk supplies. J. Dairy Research **21**, 337 (1954).
STRANDSKOV, F. B., and J. B. BOCKELMANN: Antibiotics as inhibitors of microbiological contamination in beer. J. Agr. Food Chem. **1**, 1219 (1953).
— — Inhibiting microbiological growth in brewery yeast fermentations. U.S. Pat. No. 2,964,406 (1960).
—, J. A. BRESCIA, and J. B. BOCKELMANN: The application of selective bacteriostasis of brewery bacteriology. Wallerstein Lab. Communs. **16**, 31 (1953).
STÜBER, O., and Z. BINDER: Antibiotic effect of nisin producing lactic acid bacteria. I. 15th Internat. Dairy Congress, London **2**, 552 (1959 a).
— — Antibiotic effect of nisin-producing lactic acid bacteria. II. 15th Internat. Dairy Congress, London **2**, 562 (1959 b).
STUMBO, C. R., L. VORIS, B. G. SKAGGS, and B. HEINEMANN: A procedure for assaying residual nisin in food products. J. Food Sci. **29**, 859 (1964).
TACQUET, A., F. TISON, and B. DEVULDER: Bactericidal action of yoghurt against mycobacteria. Ann. Inst. Pasteur. **100**, 581 (1961).
TAKAHASHI, Y.: Chlortetracycline-containing food preservative. Japanese Pat. No. 15,641 (1962).
TAKATSUKA, N., T. TAWARA, and H. ONO: Efficacy of some antibiotics as preservatives for fresh tunny. Bull. Japan Soc. Sci. Fisheries **23**, 221 (1958).
TARR, H. L. A.: Antibiotics as a preservative measure. Presented June 12, 1959 at Nat. Conf., Can. Inst. Food Technol., Toronto 1959.
— Antibiotics in fish preservation. Bull. Fisheries Research Board, Can. No. **124**, 24 pp. (1960).
— Comparative value of germicidal ices for fish preservation. J. Fisheries Research Board Can. **7**, 155 (1948).
— Preservation by antibiotics and by penetrating radiations. I. Fishing News Internat. **1**, 17 (1961).
— Preservation by antibiotics and penetrating radiations. II. Fishing News Internat. **1**, 46 (1962).
—, and C. P. DEAS: Action of sulfa compounds, antibiotics and nitrite on growth of bacteria in fish flesh. J. Fisheries Research Board Can. **7**, 221 (1948).
—, and D. C. GILLESPIE: Method of producing block ice. Can. Pat. No. 552,878 (1958).
—, J. W. BOYD, and H. M. BISSETT: Experimental preservation of flesh foods with antibiotics. Abstr. Amer. Chem. Soc. Meeting, p. 30 A, Sept. 1953.
—, J. W. BOYD, and H. M. BISSETT: Experimental preservation of fish and beef with antibiotics. J. Agr. Food Chem. **2**, 372 (1954).

TARR, H. L. A., B. A. SOUTHCOTT, and H. M. BISSETT: Effect of several antibiotics and food preservatives in retarding bacterial spoilage in fish. Progress Rept. Fisheries Research Board of Canada Pacific Coast Station, No. 83, 35 (1950).

TAYLOR, J. I., A. HIRSCH, and A. T. R. MATTICK: The treatment of bovine streptococcal and staphylococcal mastitis with nisin. Vet. Record **61**, 197 (1949).

TEIXEIRA, G. C., and R. B. SCOTT: Further clinical and laboratory studies with novobiocin. I. Treatment of staphylococcal infections in infancy and childhood. II. Novobiocin concentrations in the blood of newborn infants and in the breast milk of lactating mothers. Antibiotic Med. **5**, 577 (1958).

TEJERINA, G., and A. PORTOLE'S: Preservation of milk with chlortetracycline. Ann. Bromatol. Madr. **12**, 117 (1960).

TEPLY, M.: Estimation of nisin by a dilution method using lactobacilli. Prumysl Protravin **12**, 547 (1961).

TESTA, C., and I. MORELLI: Observations on the action of tetracyclines in the preservation of Adriatic codfish. V. Residual antibiotic in the fish. Atti. Soc. Ital. Sci. Vet. **12**, 469 (1958).

THATCHER, F. S., and A. LOTT: Comparative microflora of chlortetracycline-treated and nontreated poultry with special reference to public health aspects. Applied Microbiol. **9**, 39 (1961).

THOMAS, H.: Tests on human milk preserved with streptomycin and citrates in new-born children. Mschr. Kinderheilk. **100**, 457 (1953).

THOMPSON, B. D.: Post harvest treatments of strawberries and radishes with cobalt-60 irradiation. Proc. Florida State Hort. Soc. **72**, 114 (1959).

THORNLEY, M. J., M. INGRAM, and E. M. BARNES: Effects of antibiotics and irradiation on the *Pseudomonas-Achromobacter* flora of chilled poultry. J. Applied Bacteriol. **23**, 487 (1960).

THORPE, R. H.: The action of nisin on spoilage bacteria. I. The effect of nisin on the heat resistance of *Bacillus stearothermophilus* spores. J. Applied Bacteriol. **23**, 136 (1960).

TOMIC-KAROVIC, K.: Detoxicating effect of acidophilus milk on streptomycin. Bull. Sci., Yugoslavia **79**, 166 (1962).

—, and J. FANJEK: Acidophilus milk in the therapy of infantile diarrhea caused by pathogenic *E. coli* Amm. paediat. **199**, 625 (1962).

—, D. SUDEC, and C. TOMLJENOVIC: Antibiotic action of *Lactobacillus acidophilus* upon *M. tuberculosis*. Bull. Sci., Yugoslavia **5**, 16 (1959).

TOMIYAMA, T.: Fish preservation by chlortetracycline (aureomycin). 9th Proc. Pacific Sci. Congress, Pacific Sci. Assoc., 1957 Bangkok, Thailand **5**, 128 (1963).

—, and A. MATSUDA: Effectiveness of chlortetracycline and oxytetracycline at above 5 p.p.m. level in keeping the quality of ground fish flesh. Nippon Suisan Gakkaishi **28**, 593 (1962).

—, and T. SUGITANI: Effectiveness of tetracycline treatment on quality improvement of processed foods. I. Effect of chlortetracycline treatment on quality of salt sardine and salt mackerel, and change in residual chlortetracycline during salting. Bull. Japan Soc. Sci. Fisheries **24**, 581 (1958).

—, and Y. YONE: Effect of dipping in iced chlortetracycline (CTC) containing sea water on keeping quality of mackerel aboard ship and determination of CTC residue in the fish. Nippon Suisan Gakkaishi **25**, 290 (1959 a).

— — Effects of the tetracycline treatments on the quality improvement of processed foods. II. Quality improvement and chlortetracycline residue of salted salmon. Nippon Suisan Gakkaishi **25**, 67 (1959 b).

— —, and A. MATSUDA: A comparative study on the effect of chlortetracycline and oxytetracycline below 6 p.p.m. level in keeping the quality of ground flesh of fishes. Nippon Suisan Gakkaishi **28**, 522 (1962).

— —, and A. TSUDA: The chlortetracycline (CTC) content of the CTC treated fish and of kamaboko made from the treated fish. Nippon Suisangaku Kaishi **23**, 568 (1957 a).

TOMIYAMA, T., Y. YONE, and A. TSUDA: The chlortetracycline (CTC) content of trawl catch treated on board, and the residual CTC in edible part on cooking and processing of treated fishes. Nippon Suisangaku Kaishi **23**, 572 (1957 b).

— —, and K. KOBAYASHI: Penetration of chlortetracycline (CTC) into tissues of sock-eye salmon on CTC dip treatment and its destruction on heating. Nippon Suisan Gakkaishi **24**, 1012 (1958).

— —, and K. MIKAJIRI: Penetration of chlortetracycline into fish flesh or round fish and its heat inactivation. Bull. Japan Soc. Sci. Fisheries **22**, 778 (1957 c).

— — — Uptake of aureomycin chlortetracycline by fish and its heat inactivation. Food Technol. **11**, 290 (1957 d).

— —, and S. FUJINO: On chlortetracycline residue in dip treated fish. Bull. Japan Soc. Sci. Fisheries **27**, 713 (1961).

— —, S. KUROKI, M. NOMURA, S. HARA, T. SCHICHIZI, T. MAEDA, I. ODA, and T. MATSUO: Effectiveness of chlorotetracycline on keeping quality of several bottom fish. Bull. Japan Soc. Sci. Fisheries **22**, 128 (1956).

—, S. KUROKI, D. MAEDA, S. HAMADA, and A. HONDA: A study of the effect of Aureomycin-containing sea water and ice upon storage life of round herring. Food Technol. **10**, 215 (1956 a).

— — — — — Effectiveness of chlorotetracycline (Aureomycin) on keeping quality of Pacific round herring, *Etrumeus micropus*. Bull. Japan Soc. Sci. Fisheries **22**, 120 (1956 b).

TOYOSHIMA, H.: Comparison of antibiotics and other antiseptics in preserving foods. Hakko Kogaku Zasshi **38**, 478 (1960).

TRAMER, J., and G. G. FOWLER: Estimation of nisin in foods. J. Sci. Food Agr. **15**, 522 (1964).

TREMBATH, R. H.: Effect of penicillin in milk on starter cultures. Australian J. Dairy Technol. **5**, 143 (1950).

TSUCHIYA, Y., T. NOMURA, and M. MURASE: Influence of the treatments immediately after catching on the quality of fish meat. IV. Use of chlortetracycline ice for the preservation of saury, *Cololabis saira*. Nippon Suisan Gakkaishi **25**, 569 (1959 a).

—, Y. SATO, Y. SUZUKI, H. KUDO, T. SASAKI, and M. MURASE: Studies on the influence of treatments immediately after catching upon the quality of fish flesh. II. Further examination of treatments on keeping quality of fish. Bull. Japan Soc. Sci. Fisheries **25**, 381 (1959 b).

TSUGO, T., M. IWAIDA, A. NAGAO, Y. KOBAYASHI, K. TSURUMUTO, and K. HIRAI: Improving the keeping quality of processed cheese by use of nisin produced by a variety of *Streptococcus lactis*. Shokuhin Eiseigaku Zasshi **3**, 256 (1962).

UPHAM, S. D., F. M. CALLAHAN, and J. F. WEIDENHEIMER: Antibiotic-containing ice for preservation of foods. U. S. Pat. No. 2,766,124 (1956).

UNO, T., T. TOKUMAYA, and M. NAKAMURA: Experiment on keeping quality of marine products by using antibiotics. III. Alaska pollack. Hokusuishi Geppo **16**, 480 (1959 a).

— — —, Y. TAKAYAMA, and Y. KAWAI: Experiment on keeping quality of marine products by using antibiotics. IV. Further experiment on Alaska pollack. Hokusuishi Geppo **16**, 485 (1959 b).

VAKIL, J. R., and K. M. SHAHANI: Different pathways of lactose metabolism of *Streptococcus lactis* and their sensitivity to antibiotics. J. Dairy Sci. **45**, 655 (1962 a).

— — Lactabionate and gluconat metabolism of *Streptococcus lactis*. 8th Internat. Congress Microbiol. **B**, 11.4 (1962 b).

VALLE, E.: Antifungal factor in potato leaves. Acta Chim. Scand. **11**, 395 (1957).

VAS, K.: The use of nisin in the heat preservation of tomato products. Fruchtsaft-Ind. **8**, 73 (1963).

VAUGHN, R. H., H. NG, G. F. STEWART, C. W. NAGEL, and K. L. SIMPSON: Antibiotics in poultry-meat preservation. Development in vitro of bacterial resistance to chlortetracycline. Applied Microbiol. **8**, 27 (1960).

VERLINSKI, S.: Penicillin-resistant starter. 15th Internat. Dairy Congress, London **2**, 594 (1959).

VICKERS, H. R., L. BAGRATUNI, and S. ALEXANDER: Dermatitis caused by penicillin in milk. Lancet **1**, 351 (1958).
VIERLING, K.: Antibiotically active metabolism products and enzyme systems. Ger. Pat. No. 1,040,749 (1958 a).
— Antibiotic from propionibacterium. Brit. Pat. No. 900,640 (1958 b).
VIGUE, F.: Relative efficacy of three antibiotic combinations in bovine mastitis. J. Amer. Vet. Med. Assoc. **124**, 377 (1954).
VINCENT, J. G., R. C. VEOMETT, and R. R. RILEY: Antibacterial activity associated with *Lactobacillus acidophilus*. J. Bacteriol. **78**, 477 (1959).
VIRTANEN, A. I.: Antimikrobiell wirksame Substanzen in Kulturpflanzen. Angew. Chemie **70**, 544 (1958).
— Some aspects of factors in the maize plant with toxic effects on insect larvae. Suomen Kemistilehti **B 34**, 29 (1961 a).
— Über einen neuen Typ von Glucosiden in jungen Roggen-, Weizen- und Mais-Pflanzen. Brauwissenschaft **14**, 98 (1961 b).
—, and E. J. MATIKKALA: A new sulphur-containing amino acid in onion. III. Suomen Kemistilehti **B 31**, 191 (1958).
— — Eine neue schwefelhaltige Aminosäure in der Zwiebel. Suomen Kemistilehti **B 29**, 134 (1956).
— — Isolation of *S*-methyl- and *S*-propylcysteine sulfoxide from onion and the antibiotic activity of crushed onion. Acta Chim. Scand. **13**, 1898 (1959 a).
— — The structure and synthesis of cycloalliin isolated from *Allium cepa*. Acta Chim. Scand. **13**, 623 (1959 b).
— — New γ-glutamyl peptides in onion. *(Allium cepa)*. III. Suomen Kemistilehti **B 34**, 53 (1961).
— — New γ-glutamyl peptides in onion *(Allium cepa)*. I. γ-Glutamyl-phenylalanine and γ-glutamyl-*S*-β-carboxy-β-methyl-ethyl-cysteinyl-glycine. Suomen Kemistilehti **B 33**, 83 (1960).
—, and I. MATTILA: On the γ-glutamylpeptides in garlic *(Allium sativum)*. Suomen Kemistilehti **B 34**, 73 (1961 a).
— — γ-L-Glutamyl-*S*-allyl-L-cysteine in garlic *(Allium sativum)*. Suomen Kemistilehti **B 34**, 44 (1961 b).
—, and P. K. HIETALA: Precursors of benzoxazolinone in rye plants. I. Precursor II, the aglucone. Acta Chim. Scand. **14**, 499 (1960).
—, M. HATANAKA, and M. BERLIN: γ-L-Glutamyl-*S*-*n*-propylcystein in Knoblauch *(Allium sativum)*. Suomen Kemistilehti **B 35**, 52 (1962).
—, P. K. HIETALA, and O. WAHLROOS: Antimicrobial substances in cereals and fodder plants. Arch. Biochem. Biophys. **69**, 486 (1957).
WALDER, W. O.: Antibiotics and food. Food **21**, 100 and 142 (1952).
WALKER, E. A., and J. H. TAYLOR: The effect of chlortetracycline on eviscerated frozen chickens commercially processed in Great Britain. J. Applied Bacteriol. **23**, 155 (1960).
— —, and J. G. DAVIS: The effect of chlortetracycline on eviscerated chilled chickens commercially processed in Great Britain. J. Applied Bacteriol. **23**, 145 (1960).
WALKER, H. W., and J. C. AYRES: Antibiotic residuals and microbial resistance on poultry treated with tetracyclines. Food Research **23**, 525 (1958).
— — Characteristics of yeasts isolated from processed poultry and the influence of tetracyclines on their growth. Applied Microbiol. **7**, 251 (1959).
WALTON, L., M. HERBOLD, and C. C. LINDEGREN: Bactericidal effects of vapors from crushed garlic. Food Research **1**, 163 (1936).
WEISER, H. H., H. S. GOLDBERG, V. R. CAHILL, L. E. KUNKLE, and F. E. DEATHERAGE: Observations on fresh meat processed by the infusion of antibiotics. Food Technol. **7**, 495 (1953).
—, L. E. KUNKLE, and F. E. DEATHERAGE: Use of antibiotics in meat processing. Applied Microbiol. **2**, 88 (1954).
WELCH, H.: Problems of antibiotics in food as the Food and Drug Administration sees them. Amer. J. Public Health **47**, 701 (1957).
—, W. R. JESTER, and J. M. BURTON: Antibiotics in fluid milk. Antibiotic Chemother. **5**, 571 (1955).

WELLS, F. E., and W. J. STADELMAN: Development of yeasts on antibiotic-treated poultry. Applied Microbiol. **6,** 420 (1958).

—, J. V. SPENCER, and W. J. STADELMAN: Effect of packaging materials and techniques on shelf life of fresh poultry meat. Food Technol. **12,** 425 (1958).

—, S. E. HARTSELL, and W. J. STADELMAN: Growth of psychrophiles. I. Lipid changes in relation to growth-temperature reductions. J. Food Sci. **28,** 140 (1963 a).

—, W. J. STADELMAN, and S. E. HARTSELL: Growth of psychrophiles. II. Growth of poultry meat spoilage bacteria and some effects of chlortetracycline. J. Food Sci. **28,** 254 (1963 b).

—, J. L. FRY, W. W. MARION, and W. J. STADELMAN: Relative efficacy of three tetracyclines with poultry meats. Food Technol. **11,** 656 (1957),

WHEATER, D. M., A. HIRSCH, and A. T. R. MATTICK: Lactobacillin, an antibiotic from lactobacilli. Nature **168,** 659 (1951).

WHEATON, E., and G. L. HAYS: Antibiotics and the control of spoilage in canned foods. Food Technol. **18,** 147 (1964).

—, J. D. BURROUGHS, and G. L. HAYS: Flat sour spoilage of tomato juice and its control with subtilin. Food Technol. **11,** 286 (1957).

WHITE, J. W., JR., and M. H. SUBERS: Studies on honey inhibine. 2. A chemical assay. J. Apicult. Research **2,** 93 (1963).

— — Studies on honey inhibine. 3. Effect of heat. J. Apicult. Research **3,** 45 (1964 a).

— — Studies on honey inhibine. 4. Destruction of the peroxide accumulation system by light. J. Food Sci. **29,** 819 (1964 b).

— —, and A. I. SCHEPARTZ: The identification of inhibine, the antibacterial factor in honey, as hydrogen peroxide and its origin in a honey glucose-oxidase system. Biochim. Biophys. Acta **73,** 57 (1963).

WHITEHEAD, H. R.: A substance inhibiting bacterial growth, produced by certain strains of lactic streptococci. Biochem. J. **27,** 1793 (1933).

—, and G. A. COX: The detection of penicillin in milk. J. Applied Bacteriol. **19,** 247 (1956).

—, and D. J. LANE: The influence of penicillin on the manufacture and ripening of Cheddar cheese. J. Dairy Research **23,** 355 (1956).

WILKOWSKE, H. H., and W. A. KRIENKE: Assay of various mold-ripened cheeses for antibiotic activity. J. Dairy Sci. **37,** 1184 (1954).

— — Influence of penicillin in milk on total and coliform bacterial plate counts. J. Milk Food Technol. **14,** 92 (1951 a).

— — Influence of penicillin on the lactic acid production of certain lactobacilli. J. Dairy Sci. **34,** 1030 (1951 b).

WILLIAMS, B. E.: Preserving fresh meats. U. S. Pat. No. 2,942,986 (1960).

— Tenderizing and preserving meat. U. S. Pat. No. 3,076,712 (1963).

WILLIAMS, O. B., and L. L. CAMPBELL, JR.: The effect of subtilin on thermophilic flat sour bacteria. Food Research **16,** 347 (1951).

WILSON, D. C., and H. D. BROWN: Heat-induced inhibitory agents obtained from processed fruits and vegetables. Food Technol. **7,** 250 (1953).

WILSON, G. D., P. D. BROWN, W. R. CHESBRO, B. GINGER, and C. E. WEIR: The use of antibiotics and gamma irradiation in the aging of steaks at high temperatures. Food Technol. **14,** 143 (1960).

WINKLER, S.: The inhibitory action of lactobacilli on propionic acid bacteria. 13th Internat. Dairy Congress, The Hague **3,** 1164 (1953).

—, and K. MATHA: Study of the production of nisin by lactic acid streptococci of different origin. Milchwissensch. Ber. **6,** 71 (1956).

—, and M. FRÖHLICK: The effect of nisin on the microorganisms most important to the ripening of Emmental cheese. Milchwissensch. Ber. **7,** 125 (1957).

— — Investigations of Emmental cheese with nisin. Milchwissensch. Ber. **8,** 87 (1958).

WINTER, A. G., and L. WILLEKE: Antibiotics from higher plants. V and VI. Naturwissensch. **39,** 190 and 236 (1952).

WINTERBOTTOM, R., H. P. BROQUIST, and A. R. WHITEHILL: Improving the storability of plucked poultry and poultry meat. Ger. Pat. No. 1,073,844 (1960 a).
— — — Preserving dressed poultry. U. S. Pat. No. 2,930,702 (1960 b).
WODSAK, W.: Possible uses of nisin (in preservation). Deut. Lebensm.-Rundschau **58**, 1935 (1962).
WRENSHALL, C. L., and J. R. MCMAHAN: How newly OK'd antibiotic boosts poultry shelf life. Food Eng. **28**, 53 (1956).
— —, and R. C. OTTKE: Preserving fresh meat. U. S. Pat. No. 2,942,982 (1960).
YACOWITZ, H., F. PANSY, S. WIND, H. STANDER, H. L. SASSAMAN, J. R. PAGANO, and W. H. TREJO: Use of nystatin (mycostatin) to retard yeast growth on chlortetracycline-treated chicken meat. Poultry Sci. **36**, 843 (1957).
YONE, Y., S. FUJINO, and T. TOMIYAMA: Stability and bacteriostatic effect of chlortetracycline in sea water. Nippon Suisan Gakkaishi **26**, 514 (1960).
—, and T. TOMIYAMA: Improvement of the quality of yellowfin tuna by chlortetracycline treatment aboard ship. Nippon Suisan Gakkaishi **26**, 49 (1960).
— —, and S. HAMADA: Comparison of the effect of various chlortetracycline (CTC) treatments in keeping quality of trawled fish and in CTC residue distribution in the treated fish. Nippon Suisan Gakkaishi **25**, 156 (1959).
ZIEGLER, F., and W. J. STADELMAN: The effect of aureomycin treatment on the shelf life of fresh poultry meat. Food Technol. **9**, 107 (1955).
ZIMMERMANN, M. C.: Chronic penicillin urticaria from dairy products, proved by penicillinase cures. Arch. Dermatol. (N. Y.) **79**, 1 (1959).

Anwendung und Eigenschaften des 3-Amino-1,2,4-triazols im Hinblick auf seine Rückstände in Lebensmitteln

Von

ERNST KRÖLLER*

Inhalt

I. Einführung

1888 beschreiben CURTIUS und LANG Darstellung und Eigenschaften des 3-Amino-1,2,4-triazols. Seine wachstumsregulierende und herbicide Wirkung beobachtete 1955 ALLEN von der Firma Amchem Products Inc., die sich darauf das U.S. Patent 2,670,282 erteilen ließ. Die Registrierung als Unkrautvertilgungsmittel (Herbicid) durch das *U.S. Department of Agriculture* erfolgte 1957, die Aufnahme in das Pflanzenschutzmittel-Verzeichnis der Bundesrepublik Deutschland 1959. Als sehr wirksames und selektives Herbicid fand es schnell ein weites Anwendungsgebiet im Acker- und Gartenbau, in Obstplantagen und Weinbergen.

---

* Wissenschaftlicher Oberrat im Bundesgesundheitsamt, Max-von-Pettenkofer-Institut, Abt. Lebensmittelchemie, Berlin-Dahlem.

II. Chemie

3-Amino-1,2,4-triazol, $C_2N_4H_4$ (3-AT), wird u. a. durch alkalische Zersetzung von Formylaminoguanidinsalzen, durch thermische Spaltung von 5-Amino-1,2,4-triazol-carbonsäure-3 oder durch Kondensation von Ameisensäure und Aminoguanidin erhalten (BEILSTEIN 1937, POTTS 1961). Es tritt in zwei desmotropen Formen auf, die leicht ineinander übergehen und im gegenseitigen Gleichgewicht stehen:

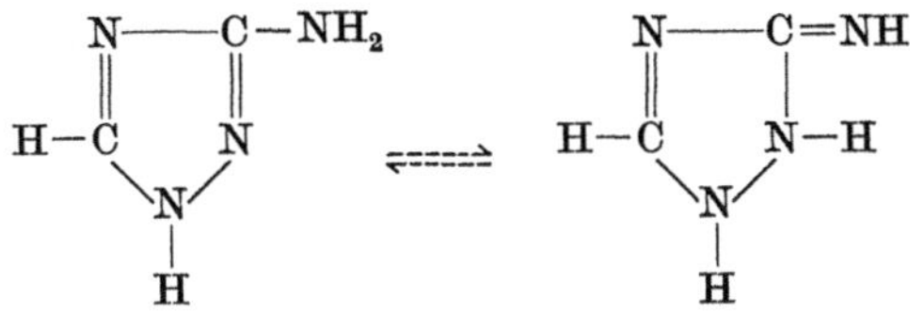

3-Amino-1,2,4-triazol (3-AT)   1,2,4-Triazolon-3-imid

Dies zeigt sich z. B. deutlich beim Lösen des 3-AT in Wasser. Führt man unmittelbar danach eine quantitative Bestimmung über seine Diazoverbindung aus, so werden etwa nur 75 Prozent 3-AT erfaßt. Beim Stehen der Lösung verschiebt sich dann das Gleichgewicht langsam zur Seite der Aminform, so daß z. B. nach einer Stunde etwa 85 Prozent, nach zwei Stunden etwa 95 Prozent und nach fünf Stunden 100 Prozent darin vorliegen (KRÖLLER 1961). Das reine 3-AT kristallisiert in schwach gelblichen feinen geruchlosen Nadeln von bitterem Geschmack, die bei 195° C. schmelzen und bei höherer Temperatur unter teilweiser Zersetzung sublimieren, Molekular-Gewicht = 84,08. Die Löslichkeit in 100 ml. Wasser beträgt rd. 28 g. bei 20° C. bzw. 52 g. bei 50° C. Äthanol löst bei 75° C. 26 g. in 100 ml. In Chloroform, Methylenchlorid, und Acetonitril ist die Löslichkeit des 3-AT geringer, sehr gering in Essigsäureäthylester und praktisch gleich null in Diäthyläther, Kohlenwasserstoffen und Aceton. Die wäßrigen Lösungen reagieren lackmusneutral. Es besitzt basische Eigenschaften und verbindet sich mit Säuren zu Salzen wie u. a. dem Hydrochlorid (Sp. 153° C.), Nitrat (Sp. 174° C.) und Pikrat (Sp. 227° C.). Die Alkalimetalle bilden mit 3-AT amidartige, leicht hydrolysierbare Verbindungen. Glucose verbindet sich damit in schwach saurer wäßriger oder methanolischer Lösung zu einem in weißen Kristallen ausfallenden Aminglucosid vom Sp. ~ 315° C., das weder mit *p*-Anisidin, noch mit Ninhydrin eine Reaktion zeigt (HERRETT und BAGLEY 1964) und dessen Infrarotspektrum FREDRICK und GENTILE

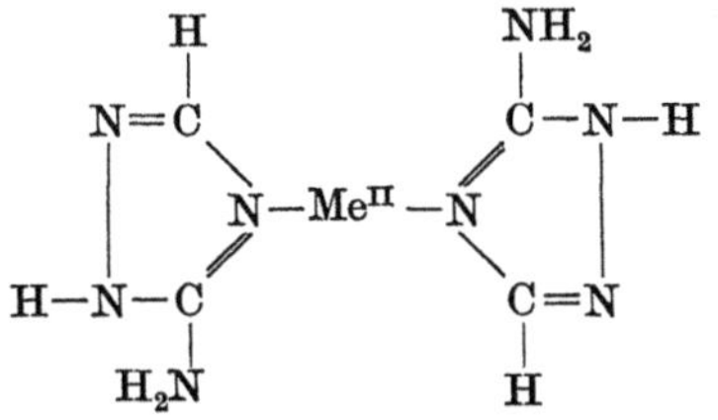

(1961) untersuchten. Mit Aldehyden und Ketonen erfolgt Reaktion zu Aldimiden bzw. Ketimiden, und mit vielen zwei- und dreiwertigen Metallionen entstehen stabile Chelatkomplexe vom Typ vorstehender Formel.

Von diesen Verbindungen sind besonders die des Co, Ni, Fe^{II}, Cd, Zn, Hg^{II} und Ag schwer löslich und fallen beim Zusammengeben wäßriger Lösungen des Metallsalzes und des 3-AT als flockige leicht absitzende Niederschläge aus, wobei sich vor allem die weißen Ag- und Hg^{II}-Verbindungen durch besondere Schwerlöslichkeit auszeichnen. Die Grenzkonzentration einer 3-AT-Lösung, bei der noch Trübungen auftreten, liegt für die Silberverbindung bei 0,01 Prozent und für die Quecksilberverbindung bei 0,007 Prozent. Die ausfallenden Komplexe sind in Diäthyläther löslich, wenig löslich in Methylenchlorid, und unlöslich in Tetrachlorkohlenstoff. Interessant ist das Verhalten des rosa ausfallenden und beim Entwässern mit Aceton blau werdenden Cobalt-II-Komplexes. Dieser geht an der Luft, besonders leicht in Gegenwart von Wasser, unter Braunfärbung in die Cobalt-III-Verbindung über, deren Lösung mit Natronlauge ein klares braunes Gel liefert.

Keine Fällung, jedoch Farbänderung infolge Komplexbildung wurde beobachtet bei Kupfer-II-salzen (kräftige Grünfärbung der Lösung durch 3-AT, die bei Zugabe von Natronlauge in blauviolett übergeht), sowie bei Titan-III-salzen (tief blaue Färbung) (Kröller 1961). Auch Komplexe mit Mg und Mn sind beschrieben worden (Sund 1956, Fredrick und Gentile 1960 a). 3-AT ist gegen Luft und Licht recht beständig, so daß eine wäßrige bei 30° C offen aufbewahrte Lösung nach acht Tagen keine Veränderung zeigt. Auch beim Eindampfen mit 2*N*-Ammoniak, fünfprozentiger Trichloressigsäurelösung oder 1*N*-Salz- oder Schwefelsäure wird es nicht zerstört. Während heißes verdünntes Wasserstoffperoxid ohne Wirkung bleibt, wird 3-AT von starken Oxydationsmitteln zu 3,3'-Azo-1,2,4-triazol oxydiert. Mit Nitriten entsteht in Gegenwart von konzentrierter Salzsäure 3-Chlor-1,2,4-triazol, mit verdünnter Salzsäure dagegen ein beständiges Diazoniumsalz, das u. a. mit $AuCl_3$ eine unlösliche, in gelben Kristallen anfallende Verbindung bildet und mit Phenolen, Aminen usw. zu Farbstoffen kuppelt. Ammoniakalische Silbernitratlösung wird von 3-AT in der Siedehitze langsam reduziert, Fehligsche Lösung dagegen nicht. Wegen seiner stark polaren Eigenschaften wird es aus seinen Lösungen an Ionenaustauschern abgeschieden und kann so von Begleitstoffen getrennt werden. Auch von oberflächenaktiven Bodenarten wird das 3-AT meist über längere Zeiträume festgehalten. Seine Lösungen zeigen sowohl im sichtbaren als auch im UV-wellenlängenbereich zwischen 200 und 650 nm kein Absorptionsmaximum.

III. Toxizität

3-AT besitzt ähnlich dem Thiouracil antithyroide Wirksamkeit, d. h. es verhindert die Aufnahme von anorganisch gebundenem Jod durch die Schilddrüse und damit die Bildung des Thyroidhormons (Jukes und Shaffer 1960, Alexander 1959 a und b). Die akute orale LD_{50} beträgt für Mäuse 14,7 g./kg. und für Ratten (männlich Albino) 25 g./kg. (Durham 1961). Die intravenöse Anwendung bei Mäusen (1,6 g./kg.), Katzen (1,72 g./kg.), und Hunden (1,2 g./kg.) erbrachte keine Anhaltspunkte für eine stark toxische Wirkung des 3-AT. Ebenso wurde eine einmalige intraperitoneale Injektion von 4 g./kg. 3-AT von Mäusen gut vertragen (SHAFFER 1956). Im subakuten Toxizitätsversuch mit 1000 und 10 000 p.p.m. 3-AT in der Gesamtnahrung wurde bei Ratten nach 63 Tagen ein Zurückbleiben der Gewichtszunahme und fettige Degeneration der Leberzellen festgestellt. Der chronische Toxizitätsversuch bestätigte die vorwiegend stark antithyroidale Wirkung, wobei die Schwelle dieser Wirksamkeit (Kropfbildung) bei etwa 50 p.p.m. 3-AT im Futter entsprechend einer täglichen Gabe von 2,5 mg./kg. Ratte ermittelt wurde. Bei Gaben unterhalb dieser Menge wird keine Reaktion beobachtet (EGGERT 1959), dagegen treten bei 100 und 500 p.p.m. 3-AT Wachstumsverzögerung, Schilddrüsenvergrößerung und veränderte Nahrungsverwertungsverhältnisse auf. Die mikroskopische Untersuchung ergab nichtfunktionelle Hyperplasie und toxische Degeneration der Schilddrüse, die einer partiellen Exstirpation derselben beim lebenden Tier entsprechen (WEIR *et al.* 1958). Daneben zeigten die mit 500 p.p.m. 3-AT gefütterten Ratten stark hervortretende Augen. Alle diese Erscheinungen treten bei männlichen Tieren gegenüber den weiblichen besonders ausgeprägt hervor, bilden sich jedoch beim Absetzen der 3-AT-Fütterung innerhalb von etwa zwei Wochen wieder völlig zurück.

Aus den beschriebenen Befunden kann gefolgert werden, daß 3-AT ein Kropfbildner (Goitrogen) ist, wenn es wiederholt angewendet wird. Eine Reaktion kann aber nur beim Überschreiten einer Grenzkonzentration von 3-AT in der Nahrung (bei Ratten ca. 50 p.p.m.) diagnostiziert werden, ist jedoch bei Entzug des 3-AT schnell wieder verschwunden. Nach zweijähriger Fütterung von Ratten mit 100 p.p.m. 3-AT täglich wurden bei einigen Tieren als Adenocarcinom interpretierte Schilddrüsenveränderungen festgestellt, Metastasen wurden jedoch keine gefunden (JUKES und SHAFFER 1960). Aus diesem Grunde sowie aus der Feststellung heraus, daß das Wirkungsergebnis vorwiegend von Umwandlungsprodukten innerhalb der Organe abhängt, bezeichnet OSER (1962) das 3-AT als „indirektes sekundäres Cancerogen". ALEXANDER (1959 a und b) glaubt, daß alle pathologischen Veränderungen der Schilddrüse durch das Aufheben der Aktivität der Thyroid-Enzyme Catalase und Peroxydase durch die Hemmwirkung des 3-AT auf die Jodaufnahme hervorgerufen werden. Diese Annahme stützt sich auf die Untersuchungen von HEIM *et al.* (1956), die eine Abnahme der Peroxydaseaktivität der Ratte drei Stunden nach Injektion von 3-AT und eine Normalisierung derselben nach 24 Stunden feststellten. Auch die Porphyrin-haltige Leber- und Nierencatalase sowie die an der Porphyrinsynthese beteiligte δ-Aminolävulinsäure-dehydrase werden in noch ungeklärter Weise durch 3-AT gehemmt (HEIM *et al.* 1955). Nach Versuchen von HOSHINO (1960) wird bei Ratten die Entwicklung eines experimentell mit 4-Dimethylaminoazobenzol erzeugten Leberkrebses von 3-AT signifikant gehemmt. Haut und Augen der Versuchstiere werden von 3-AT nicht irritiert (SHAFFER 1960).

Trotz der geringen Toxizität wurde für das 3-AT wegen seiner goitrogenen Wirkung und des Verdachtes cancerogener Eigenschaften vom *U.S. Secretary of Health, Education, and Welfare* der Toleranzwert für Restmengen in Lebensmitteln auf 0 p.p.m. festgesetzt (FLEMMING 1959, DURHAM

1961). Infolgedessen wurden u. a. 1957 rd. 3500 Pfund Preiselbeeren, die 3-AT-rückstände enthielten, auf Veranlassung der *U.S. Food and Drug Administration* vom Markt entfernt und vernichtet. Auch 1959 kam es zu erneuten Beschlagnahmeaktionen mehrerer Tonnen 3-AT-verseuchter Preiselbeeren und Preiselbeerkonserven (Du Shane 1959). Gleichzeitig wurden öffentliche Warnungen herausgegeben und veranlaßt, daß in allen Anbaugebieten Proben gezogen und auf 3-AT untersucht wurden (Flemming 1959, *Hazleton Laboratories* 1959).

Auch in der Bundesrepublik Deutschland dürfen, wie auch in den meisten anderen Ländern nach der in Vorbereitung befindlichen Restmengen-VO keine Rückstände von 3-AT in oder auf Obst und Gemüse nachweisbar sein.

IV. Phytotoxische Wirkung

Das 3-AT dringt als polare Substanz in Abhängigkeit von Feuchtigkeit und Temperatur der umgebenden Luft in wäßriger Lösung auf dem „Wasserweg" durch das Parenchymgewebe der Blätter und Wurzeln in das Phloem ein, von dem es dann weiter wahrscheinlich durch den Protoplasmastrom aktiviert, durch Zelle-zu-Zelle-diffusion im Mesophyll in das Xylem übergeht. Hierbei ist nach Herrett und Link (1961 b) seine Eindringgeschwindigkeit von der jeweiligen Empfindlichkeit der Pflanze abhängig, indem sie mit deren Anstieg zunimmt. Da 3-AT auf Grund seiner Struktur zum Unterschied von vielen anderen Herbiciden nur geringfügig von der lebenden Zelle festgehalten wird, wandert es frei in der wäßrigen Phase des Säftestromes zusammen mit den Assimilationsprodukten durch die gesamte Pflanze und wird, da es im Stoffwechsel keine Verwendung findet, zusammen mit diesen in den Wachstumsspitzen abgelagert (Crafts 1956 und 1959, Crafts und Yamaguchi 1958, Massini 1958, Radwan *et al.* 1960, Crafts 1961 a, Leonard und Lider 1961). Durch Herabsetzen der Oberflächenspannung des Wassers mit geeigneten Netzmitteln läßt sich die Eindringgeschwindigkeit des 3-AT in die Pflanzenzelle beträchtlich erhöhen (Jansen *et al.* 1961). Nach Freed und Montgomery (1958) erwies sich ein Alkyl-aryl-polyoxyäthylenglykol, bei dessen Anwendung in 0,05prozentiger Lösung innerhalb der ersten 24 Stunden bereits 78 Prozent des 3-AT absorbiert wurden, am wirksamsten. Das so eingedrungene, wie ein Wachstumsregulator in der gesamten Pflanze verteilte 3-AT wird während der Lichteinwirkung mit dem Strom der Photosyntheseprodukte aus den Blättern in die Wurzelspitzen oder Knollen transportiert, während es bei Dunkelheit durch den Transpirationsstrom in die Blattspitzen gedrückt wird und sich dort ansammelt. Damit kann es seine charakteristische biologische Aktivität, für die nach Schweizer und Rogers (1964 b) in erster Linie die am Triazol-Kern gebundene Aminogruppe verantwortlich ist, in allen Teilen der Pflanze besonders in ihren Wachstumszentren voll entfalten (Andersen 1958). Diese Wirkung beruht auf den Eigenschaften des 3-AT, in eine Reihe lebenswichtiger Reaktionen der Zelle einzugreifen, indem es u. a. ihre Fähigkeit aufhebt mit Systemen zu reagieren, die in wäßriger Phase freie Radikale bilden. Den Grund hierfür sehen Castelfranco und Brown

(1963) darin, daß das 3-AT unter diesen Bedingungen selbst in ein Radikal

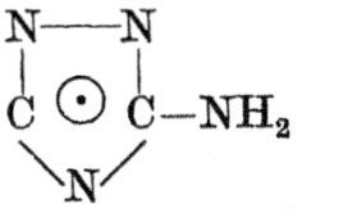

übergeht, das durch die Wechselwirkung der unpaarigen Elektronen der drei heterocyclischen *N*-atome stabilisiert ist. Dieses Radikal ist dann zur Bildung der verschiedensten Reaktions- und Umwandlungsprodukte befähigt. So werden u. a. mehrere stoffwechselregulierende Enzyme gehemmt oder ganz blockiert. PALMER und PORTER (1959) fanden z. B. bei ihren Experimenten mit *Cyperus rotundus* L., daß die Aktivität der Ascorbinsäureoxydase etwa proportional zum Anstieg der 3-AT-konzentration in der Zelle abnimmt. Auch die Catecholase wird merklich gehemmt und reichert sich in 3-AT behandelten Knollen mehr an als in unbehandelten. Während die Aktivität der DOPA-oxydase, Thyrosinase und Xanthinoxydase (HOFSTEE 1955) von geringen 3-AT-mengen gehemmt, von hohen Konzentrationen völlig blockiert wird, hängt der Grad der Peroxydasehemmung weitgehend von der Art ihrer prosthetischen Gruppe ab. Auch die Catalaseaktivität wird von 3-AT je nach Pflanzenart beeinflußt (bei Kartoffeln, Sonnenblume und Gerste beispielsweise deutlich weniger als bei Mais oder Lupine). Dabei ist nach PYFROM *et al.* (1957) die in vitro-Hemmung von der Temperatur abhängig. Die Inaktivierung erfolgt irreversibel durch Reaktion zwischen dem Catalase-H_2O_2-komplex und dem 3-AT, wobei dieses in ein hochaktives Radikal übergeht, das alkylierend auf das Enzym-Protein einwirkt und es so blockiert (MARGOLIASH und NOVOGRODSKY 1958, MARGOLIASH *et al.* 1960, INGRAHAM 1962). Daneben spielt die Komplexbildung des Catalase-eisenatoms mit 3-AT ebenfalls eine Rolle für die Hemmwirkung. Beim Verschwinden des 3-AT aus der Zelle kehrt die ursprüngliche Catalase-Aktivität schnell wieder zurück. Zu entsprechenden Ergebnissen führten auch die Arbeiten von FEINSTEIN *et al.* (1958). Die Aktivität der Fettsäure-Peroxydase wird in ähnlicher Weise beeinflußt (CASTELFRANCO 1960). Durch unterschwellige Dosen geschädigtes Gewebe zeigt eine höhere Ascorbinsäureoxydase-Aktivität als unbehandeltes (McWHORTER 1958), wie überhaupt durch 3-AT die Atmungs- und Oxydationsvorgänge des pflanzlichen Organismus aktiviert werden (McWHORTER und PORTER 1958). Dabei besteht eine Abhängigkeit der Wirksamkeit der die Atmung steuernden Oxydationsenzyme von den der 3-AT-behandlung vorausgegangenen Wachstumsbedingungen und dem Alter der Pflanze, sowie von der verwendeten 3-AT-menge und seiner Bindungsform (McWHORTER und PORTER 1960, HERBERT und LINK 1957). Bei Baumwollpflanzen wird die Atmungstätigkeit etwa 26 Stunden nach der 3-AT-behandlung wieder normal (MILLER und HALL 1957 a).

Neben diesen Wechselwirkungen mit den Enzymen erstreckt sich die Toxizität des 3-AT auch auf eine Hemmung der Chlorophyllbildung, wodurch viele Pflanzen absterben. Diese beruht nach Untersuchungen von PYFROM *et al.* (1957), ROGERS (1957 a) und MILLER und HALL (1957 b) nicht auf einer chemischen Veränderung des bereits in der Pflanze vorhandenen Chlorophylls oder auf einem Einbau von 3-AT in das Chlorophyll-Molekül bei seiner Entstehung, was auch insbesondere die Versuche von

Naylor (1964) bestätigten, der das Holochrom im Dunkeln gewachsener Pflanzen abzentrifugierte und dann im Spektralphotometer die Umwandlung zum Farbstoff verfolgte, die auch in Gegenwart von 3-AT ungestört vonstatten ging, sondern darin, daß die Neubildung des Chlorophyll-*A*-pigments — stark von der Natriumverbindung des 3-AT, schwächer von seinem Acetat — reduziert wird. Den Grund für diese Abnahme sehen Pyfrom *et al.* (1957) sowie Wolf (1960) in einer Wechselwirkung des 3-AT mit den in Entwicklung befindlichen Plastiden, wodurch die Fähigkeit des Blattes zur Pigmentbildung gehemmt wird. Auch Rogers (1957 a und c) macht die Blockierung der Chloroplastenbildung hierfür verantwortlich. Diese Hemmwirkung wird von Sund und Little (1960) mit der starken Beeinflussung der Riboflavinsynthese durch das 3-AT erklärt, während Aldrich (1958) einen Zusammenhang mit der Ribonucleinsäurebildung vermutet. Untersuchungen von Schweizer und Rogers (1964 a) zeigten jedoch, daß der Ribonucleinsäuregehalt 3-AT-behandelter Maissprosse dem der unbehandelten gleich war, dagegen war indes der säurelösliche Nucleotidanteil erheblich herabgesetzt. Wie beim Chlorophyll sind auch bei den Anthocyanidinen keine Veränderungen des Moleküls durch die Einwirkung von 3-AT festgestellt worden (Miller und Hall 1957 b). Eine wichtige Wechselwirkung zeigt das 3-AT dagegen mit den Proteinen, die mit ihm in Gegenwart freier Radikale und unter Lichteinwirkung zu Komplexen zusammentreten. Diesem als Entgiftungsreaktion der Zelle zu wertenden Vorgang läuft eine Spaltung des Komplexes im Dunkeln entgegen, die als Vergiftungsreaktion bezeichnet werden muß (Castelfranco und Brown 1963).

Als sichtbare Folge dieser Einzelwirkungen tritt etwa acht Stunden nach der 3-AT-gabe bei den Pflanzen eine an den Triebspitzen einsetzende und sich nach unten weiter ausbreitende Chlorose d. h. eine Farbaufhellung der grünen Blätter ein, die bis zu völligem Albinismus führen kann und oft mit Nekrosen des Gewebes einhergeht (Hall *et al.* 1954, Crafts 1956, Rogers 1957 a). Gleichzeitig wird das Austreiben der Wurzeln oder Knollen und das Größenwachstum der Triebe und damit der gesamten Pflanze gehemmt (Palmer und Porter 1959, Hauser und Thompson 1959, Hollingsworth und Ennis 1956). Diese einer chronischen Vergiftung entsprechenden Erscheinungen, die durch einen hohen Phosphorgehalt des Bodens erheblich intensiviert werden (Upchurch *et al.* 1963) und meist zum Absterben der Pflanze führen, können durch Spurenelemente wie z. B. Fe, Mn, Mg usw. nicht beeinflußt werden. Da das 3-AT den Einbau des Phosphors in verschiedene Nucleotide stört (Wort und Longhman 1961) lassen sich, wie Herbert und Link (1957) beobachteten, einige Ausfallerscheinungen durch Phosphatgaben merklich bessern. Auch einige Pyrimidin- und Purinderivate wie u. a. Adenin, Guanin, Hypoxanthin und Uracil, ihre Riboside Adenosin, Guanosin, Inosin sowie Riboflavin, Flavinmononucleotid und Flavinadenindinucleotid zeigen bei höheren Pflanzen (z. B. Tomaten) Antagonismus gegen 3-AT (Aldrich 1958, Sund *et al.* 1960, Castelfranco und Oppenheim 1961). Dies beruht nach Castelfranco *et al.* (1963) sowie Castelfranco und Brown (1963) beim Riboflavin auf einer Photosensibilisierung des 3-AT, wobei die vom Isoalloxazin-Nucleotid absorbierte Licht-menge als chemische Energie neben dem 3-AT auch das Flavin selbst zerstört. Da 3-AT die charakteristische Fluorescenz des Riboflavins unterdrückt, wird geschlossen, daß das aktive Isoalloxazin-Ringsystem davon inaktiviert wird. Auf diesem Effekt beruht auch die Beobachtung, daß es Chlorophyllextrakte vor photochemischer Zersetzung schützt. Es ist wahrscheinlich, daß diese Wirksamkeit auch für andere Pflanzenpigmente zutrifft. Während 1-Histidin durch 3-AT verursachte Wachstumsverzögerungen aufhebt, was die Hypothese einer Blockierung seiner Synthese durch das 3-AT stützt, lassen einfache aliphatische Aminosäuren keinerlei Wirksamkeit erkennen (HILTON 1960, JACKSON 1961). Da das 3-AT allein schon aus strukturellen Gründen keine den Auxinen ähnliche Eigenschaft aufweist, vermuten FRANS *et al.* (1957), daß seine Wachstumsbeeinflussung auf Wechselwirkung mit einigen für das Wachstum erforderlichen Inhaltsstoffen der Zellen beruht. So wird z. B. das Kohlenhydratgleichgewicht im Pflanzengewebe durch die atmungsbeschleunigende Wirkung des 3-AT infolge Oxydation zu Kohlendioxid stark verschoben (HALL *et al.* 1954). Dabei nimmt der Anteil an reduzierenden Zuckern und Saccharose innerhalb 48 Stunden um die Hälfte ab, während der ebenfalls auftretende Verlust an Stärke durch Umwandlung einer entsprechenden Menge Hemicellulose gedeckt wird. Gleichzeitig wird die normalerweise verlaufende Saccharosesynthese des Blattes aus Glucose und Fructose unterbrochen (NYLOR 1964). Auf den osmotischen Druck in der Zelle, ihre elektrische Leitfähigkeit, den pH-Wert und den Gesamtstickstoff-Gehalt sowie auf die Hygroskopizität der getrockneten Wurzeln hat das 3-AT dagegen nach Untersuchungen von WELCH und VEATCH (1962) an Quecken keinen Einfluß. Eine Erhöhung des Aschegehaltes bei den Wurzeln und Knollen 3-AT-behandelter Pflanzen fand HAUSER (1963). Auch in die Samenanlage der Pflanzen wird das 3-AT insbesondere als im Blatt gebildetes Umwandlungsprodukt geleitet und dort abgelagert (MILLER und HALL 1961). Infolge seiner toxischen Wirkungsweise verhindert oder verzögert es dann das Auskeimen der Saat (HILL *et al.* 1963). Durch Aufheben des Geotropismus zeigen die Keimwurzeln ein seitlich oder nach oben gerichtetes Wachstum und durch Entwicklungsstörungen bleiben sie kurz und dick und sind unfähig Nahrung zu resorbieren (COUNTS 1961). Nach HILTON (1962) vermag Riboflavin die Keimhemmung durch 3-AT aufzuheben.

Das gleichermaßen im Phloem als auch im Xylem der Pflanzen sich leicht und schnell bewegende und damit in alle Teile, insbesondere in ihre Wachstumszentren gelangende 3-AT besitzt mit seinen im vorhergehenden beschriebenen stark phytotoxischen Wirkungen in hervorragendem Maße alle Eigenschaften, die zum sicheren Abtöten ganzer Pflanzen erforderlich sind (ANDERSEN 1958). Dabei zeigt es seine größte Wirksamkeit zur Zeit der Reservespeicherung im Herbst, wenn die Aktivität der Wurzeln geringer ist als im Frühjahr und Sommer (CRAFTS 1959). Diese Beobachtung wurde ebenfalls von DOLL (1960) bestätigt. TIMMONS *et al.* (1958) stellten außerdem noch fest, daß keine Wirkungssteigerung des 3-AT durch Zusatz von Dieselöl oder Emulgatoren zu erzielen ist.

Untersuchungen an Kulturpflanzen ergaben, daß vor allem die Preiselbeere gegen die Wirkungen des 3-AT sehr resistent ist (ALDRICH 1956, DEmoranville und Cross 1957 a, Meggiti und Aldrich 1959) und daß Spritzungen damit vor allem im Juli und August weder Einfluß auf die Blattfarbe noch auf Ertrag und Qualität der Früchte hatten. Wird dagegen während der Hauptwachstumsperiode Mai—Juni gespritzt, so verfärben sich innerhalb sieben Tagen die Blätter proportional zur angewandten 3-AT-menge hellgelb, werden jedoch nach zwei bis drei Wochen wieder grün. Stengel, Blüten und Früchte werden nicht beeinflußt (Dana 1957, Demoranville und Cross 1957 b und 1958). Ebenfalls recht widerstandsfähig zeigen sich Stachelbeeren, Citrusarten, Weinreben, Getreidearten, Kohlrüben, Kürbisarten, Zuckerrüben, Gladiolen, verschiedene Ericaceen, sowie von den Holzgewächsen Robinien, Holunder, Eichen, Kastanien und andere. Empfindlicher, aber unbeeinflußt von einer Vorbehandlung des Bodens mit 3-AT sind Baumwolle, Kartoffeln, Mais, Zuckerrohr, Runkel- und Steckrüben, Weizen, Schoten, Bohnen und Sojabohnen. Nicht mit 3-AT dürfen behandelt werden: Luzerne, Klee und Stauden aller Art (Ennis 1956, A.C.P. 1957, Bruns und Dawson 1959, Ercegovich 1958, Crafts 1961 b und c).

V. Mikrobiologische Wirksamkeit

Nach den Arbeiten von Hilton (1960) hemmt 3-AT das Wachstum von *Saccharomyces cerevisiae* und *Schizosaccharomyces pombe* nur in höheren Konzentrationen merklich. Histidin zeigt dabei wohl vereinzelt eine Schutzwirkung, kann nach den vorliegenden Befunden jedoch nicht als Antagonist angesehen werden. Leucin, Tyrosin, Isoleucin, Methionin, Tryptophan und wahrscheinlich auch Phenylalanin heben die 3-AT-wirkung teilweise auf, während andere einfache Aminosäuren wie auch bei den höheren Pflanzen keine Wirkung zeigen. Auf das Wachstum von *Salmonella typhimurium* besitzt 3-AT keinen Einfluß und zeigt sich auch selbst in Dosen bis zu über 100 p.p.m. wirkungslos auf die Mikroflora des Bodens (Whiteside und Alexander 1960). Bei *Escherichia coli* und *Lorula cremoris* dagegen wird die Biosynthese des Adenins von 3-AT blockiert und bei *Eremothecium ashbyii* die Riboflavinsynthese gehemmt (Sund und Little 1960). Diese Wirkung wird nach Smith (1954) jedoch von Purinen wieder aufgehoben. Einen Rückgang der Phosphorylaseaktivität in Algen (*Chlorella*-Arten), der jedoch von Fe- und Mn-Ionen wieder aufgehoben wird, fanden Frederik und Gentile (1960), während Guerin-Dumartrait (1960) an ihnen eine Störung der Neubildung von Zellen durch 3-AT und eine Zellwucherung bis zum Doppelten der normalen Größe feststellten, wobei — am aufgenommenen Sauerstoff gemessen — ihre Photosynthese merklich abnimmt. Über Beeinflussungen der 3-AT-Wirkung auf Mikroorganismen durch Komplexbildung mit Schwermetallen berichtet Aaronson (1960).

VI. Verhalten im Boden

Vom Boden aufgenommenes 3-AT verliert meist schnell an Wirksamkeit (Bruns und Dawson 1959, A.C.C. 1955, A.C.P. 1955). In vitro-Versuche von Ercegovich (1958) zeigen, daß die Inaktivierung des 3-AT von der Art des Bodens sowie von der Temperatur, dem pH-Wert und seinem Feuchtigkeitsgehalt abhängt. Über- oder Unterschreiten eines pH von sieben beschleunigt ebenso wie Temperaturerhöhung und zunehmende Feuchtigkeit den Abbau des Wirkstoffes. In leichten Sandböden hält sich das 3-AT länger als in schwerem Lehmboden. Auch spielt die mikrobiologische Aktivität dabei eine bedeutende Rolle (ERCEGOVICH und FREAR 1964, JORDAN 1960). Dies beweisen u. a. auch die Untersuchungen von RIEPMA (1962), der in Böden mit hohem Anteil an organischer Substanz (z. B. Humus) einen schnelleren Abbau des 3-AT beobachtete als in solchen mit geringem Anteil, wobei die Zersetzungskurve den für das Bakterienwachstum typischen Verlauf zeigt. Sterilisierte Böden und solche, denen Gifte wie Hydroxylamin, Azide und Arsenite zugesetzt waren, verändern 3-AT nicht. Der Zersetzungsprozeß verläuft als Reaktion 1. Ordnung, so daß die Arrhenius'sche Gleichung zur Berechnung der Aktivierungsenergie angewendet werden kann, die für das 3-AT rd. 5300 Cal./Mol. beträgt (BURSCHEL und FREED 1959). Unter Berücksichtigung des Einflusses von Bodenart, pH, Temperatur und Feuchte läßt sich die Abbaugeschwindigkeit des im Boden enthaltenen 3-AT unter Feldbedingungen errechnen. Sie beträgt z. B. in einem normalen Lehmboden bei 29° C. rd. 1,3 μg. 3-AT pro g. Boden pro Tag. Damit wird das 3-AT rascher zerstört als z. B. das IPC (Isopropyl *N*-phenylcarbamat) oder CIPC [Isopropyl *N*-(3-chlorphenyl)carbamat]. SUND (1956) ermittelte eine feste Relation zwischen Abnahme der 3-AT-menge und Basenaustauschfähigkeit des Bodens, wobei sich das 3-AT ähnlich dem NH_4-Ion verhält. Daneben soll es sich auf Grund seiner Komplexbildungsfähigkeit mit zwei- oder dreiwertigen Metallen so fest chemisch binden, daß es nicht mehr nachgewiesen werden kann. Diesen Reaktionen kommt jedoch nach DAY *et al.* (1961) für die Abnahme des 3-AT im Boden nur eine untergeordnete Bedeutung zu.

VII. Umwandlungsprodukte (Metabolite)

a) Im tierischen Organismus

Fütterungsversuche mit radioaktivem C^{14}-markiertem 3-AT an weißen Ratten zeigten, daß während der ersten 24 Stunden die Hauptaktivität (70 bis 90 Prozent) im Urin auftritt. Der restliche Teil war in wechselnden Mengen in den Faeces, den Organen und im Gewebe verteilt. Die in der Atemluft nur in sehr kleinen Mengen als $C^{14}O_2$ nachweisbare Aktivität läßt darauf schließen, daß das 3-AT nur geringfügig im Organismus oxydiert wird. Von den Organen weist die Leber eine Stunde nach der Aufnahme den höchsten 3-AT-gehalt auf ($\sim$ 0,7 Prozent). Die chromatographische Trennung ihres Äthanolextraktes ergab neben unverändertem 3-AT ein Umwandlungsprodukt (Metabolit I), dessen Menge mit fortschreitender Zeit zunimmt, während gleichzeitig die des unveränderten Ausgangsstoffes abnimmt und nach 24 Stunden die Nachweisbarkeitsgrenze unterschreitet. Geringe Aktivitäten sind auch im Magen und den Eingeweiden lokalisiert, während Muskeln, Haut und Fettgewebe kein 3-AT aufweisen. Die Ausscheidung des Wirkstoffes erfolgt vorwiegend im Urin, der neben dem in der Leber gebildeten Metaboliten I, der in kleinen Mengen auch in den Faeces gefunden wurde, noch ein weiteres Umwandlungsprodukt (Metabolit II) aufweist. Während bei dem Stoff I ein 3-AT-Carboxylase-Komplex vermutet wird, dem sehr wahrscheinlich die eigentliche toxische Wirkung zukommt, ist die Zusammensetzung von II noch unbekannt (FANG *et al.* 1964).

b) Im pflanzlichen Organismus

Weit ausführlicher als bei Tieren ist die Umwandlung im Organismus der Pflanzen untersucht worden. So beschreiben u. a. MILLER und HALL (1961) ein nicht identifiziertes wasserlösliches Umwandlungsprodukt in der Saat 3-AT-behandelter Baumwollpflanzen und RACUSEN (1958) zwei aus Pflanzenteilen isolierte Verbindungen *X* und *Y* mit geringerer Toxizität als 3-AT, deren Zusammensetzung ebenfalls nicht aufgeklärt werden konnte. Die gleichen Verbindungen, sowie noch eine weitere unbekannte stärker herbicid als 3-AT wirksame Substanz „Metabolit III" beschreiben HERRET und BARGLEY (1964), während es NAYLOR (1964) auf dünnschichtchromatographischem Wege gelang, sogar über 12 Metabolite des 3-AT nachzuweisen, darunter u. a. ein bereits von ROGERS (1957 b) beschriebenes D-Glucose-3-AT-addukt im Molverhältnis 1 : 1, dessen Entstehung unter physiologischen Bedingungen in der Pflanze FREDRICK und GENTILE (1960 b) untersuchten. Dieses Aminoglucosid reagiert mit Hexocinase, inaktiviert in seiner phosphorylierten Form, die ebenfalls in 3-AT-behandelten Pflanzen vorkommt, viele Enzyme und greift so in wichtige Lebensvorgänge ein. Weitere wichtige Reaktionsprodukte des 3-AT mit Pflanzen-Inhaltsstoffen sind die Ninhydrin-positiven Verbindungen mit Glycin und Serin. Diese bilden „Zwitterionen", deren anionische Form die größte Reaktionsfähigkeit besitzt und die gleichen chemischen und toxischen Reaktionen zeigt wie das 3-AT (CARTER und NAYLOR 1959, 1960 und 1961; NAYLOR 1964). Diese Komplexe zeichnen sich vor allem durch große Widerstandsfähigkeit gegen chemische Eingriffe aus. Mit Alanin wird das von MASSINI (1959) aufgefundene 3-Amino-1,2,4-triazolyl-alanin gebildet, das aus seiner Lösung in 70prozentigem Äthanol mit Silbernitrat als schwer löslicher Niederschlag der Formel $C_5H_9N_5O_2AgNO_3$ ausgefällt wird. Weitere in Pflanzen nachgewiesene 3-AT-Komplexe enthalten als Komponente Glycolsäure oder Catalase, bei der das 3-AT an das Eisenion des Porphyrinringes fest gebunden vorliegt (MARGOLIASH *et al.* 1958 und 1960). Mit Tannin tritt das 3-AT ebenfalls zu einer nur sehr schwer spaltbaren Verbindung zusammen (KRÖLLER 1961). Ein mit Ninhydrin unter Blaufärbung reagierendes Umsetzungsprodukt ist noch nicht sicher identifiziert worden (NAYLOR 1964).

Auch in Mikroorganismen verbindet sich das 3-AT zu Metaboliten, von denen etwa 13 nicht näher identifizierte aus behandelten Böden isoliert wurden. Die Hauptmenge wird jedoch von den Bodenbakterien zu Kohlendioxid oxydiert (ASHTON 1961 und 1963, MACRAE und ALEXANDER 1965), wobei jedoch noch nicht sicher feststeht, ob es als *N*- oder *C*-Quelle zu ihrer Ernährung dient (ERCEGOVICH und FREAR 1964).

Aus Milch isolierten ALDRICH und McLANE (1957) papierchromatographisch ein Protein-3-AT-additionsprodukt.

VIII. Handelspräparate

In den Handel gelangt das 3-AT vorwiegend als etwa 50prozentige wäßrige Lösung meist mit Zusatz von Netzmitteln oder als Pulvermischung zum Stäuben unter den geschützten Namen: Amitrole, Amizol, Aminotriazol, Amino-Triazole-Weedkiller, Weedazol, Cytrol, 3 AT, ATA und anderen mehr. Auch als Bestandteil von Kombinationspräparaten ist 3-AT weit verbreitet. So vor allem als Gemisch mit Ammoniumthiocyanat unter dem Namen „Amitrol T", mit Simazin [2-Chloro-4,6-bis(äthylamino)-*s*-triazin] als „Domatol", mit Simazin und MCPA (2-Methyl-4-chlorophenoxyacetsäure) als „Domatol spezial W", mit 2,4-D und TCA (Trichloroessigsäure) als „Unkrautvernichter 2374 Bayer", mit 2,4-D und Atrazin (2-Chloro-4-äthylamino-6-isopropylamino-*s*-triazin) als „Vorox i" usw.

IX. Anwendungsart

Diese Präparate können zur Unkrautbekämpfung sowohl als Sprüh- oder Spritzmittel als auch zum Stäuben der Kulturen verwendet werden. Desgleichen sind sie als Zusatz zum Gießwasser und zum Wasser von Bächen und Teichen verwendbar.

X. Anwendungsgebiet

3-AT ist vor allem gegen Unkräuter der Pflanzenfamilien der Seggen (Riedgras, Schilfgras), der Simsen (Binsen), der Schachtelhalm- und Wolfsmilchgewächse, Farne, Compositen, Primeln (Schlüsselblumen), Rosen (Hekkenrosen usw.), Minzen und Weidengewächse sowie gegen verschiedene Cyperusarten, Wasserwegerich, Froschlöffel, Katzenschwanz (Typha-Arten), verschiedene Grasarten, Buchweizen, Sumpfklee, Fuchsschwanz, Rasenschmiele, Disteln, Rotwurzeln und viele andere wirksam (Lee 1955, Tafuro und Beatty 1955, Raleigh 1955, Pavlychenko 1955, Crafts 1957, Dana 1959, Palmer und Porter 1959, Meggiti und Aldrich 1959, Derscheid *et al.* 1960 a und b, Levi 1960). Ebenfalls werden die Zwiebeln verschiedener Alliumarten durch 3-AT zuverlässig abgetötet (Davis *et al.* 1962). Von diesen Unkräutern kommen viele in den Preiselbeerkulturen (Sümpfen) vor, aus denen sie früher nur mühsam von Hand oder mit unzureichenden chemischen Mitteln entfernt werden konnten (Cross 1949, Dana 1956). Auch zur Unkrautbekämpfung im Ackerbau (Mais-, Getreide-, Kartoffel- und Rübenfelder), auf Weiden, im Garten- und Weinbau, in Treibhäusern sowie auf Wegen und Plätzen und in Wasserläufen und Teichen dient 3-AT als wirksames Mittel (Sund *et al.* 1955, Bruns und Dawson 1959, Levi 1960, Nylund 1954, Frank *et al.* 1963). Schließlich erfreut es sich auch im Obstbau (Apfel-, Birnen- und Citrusfrucht-Plantagen) zum Beseitigen des Unkrautes zwischen den Bäumen verbreiteter Anwendung, obwohl hierbei der gleiche Effekt auch durch maschinelles Jäten leicht zu erzielen ist (Doll 1960, Leefe und Longley 1960). Nach Hall *et al.* (1953 und 1954), Alverson (1957) und Miller und Hall (1957 a) ist das 3-AT auch zum Entblättern der Preiselbeer- und Baumwollpflanzen und damit zur Erleichterung der Ernte besonders geeignet. Im Feldtest zeigte sich dabei kein Unterschied zwischen 3-AT und seinem Acetat, Hydrochlorid, Lactat, Nitrat, Phosphat sowie seiner Natriumverbindung. Dies führen Miller und Hall (1961) auf eine primäre Metabolitbildung zurück, die dann erst die Wirkung auslöst.

Trotz seiner ausgezeichneten herbiciden Eigenschaften besitzt das 3-AT nicht gegen alle Unkrautarten eine gleich starke Wirksamkeit (Yamaguchi and Crafts 1959). So werden von ihm z. B. nur etwa zehn Prozent des Johnsongrases abgetötet (Hauser und Thompson 1959), während es gegen Bahiagras und *Centaurea*-Arten praktisch wirkungslos ist (Black und Rodgers 1960, Derscheid et al. 1960 b). Auch bei submersen Wasserunkräutern ist seine Wirksamkeit nur sehr gering (Frank *et al.* 1963). In solchen Fällen werden dann an Stelle des 3-AT allein Kombinationspräparate mit anderen Herbiciden angewandt, die eine gezieltere Wirkung gegen die schwer bekämpfbaren Unkräuter aufweisen und gleichzeitig gestatten, die Anwendungsmenge des 3-AT pro Flächeneinheit beträchtlich herabzusetzen. So wird z. B. zur Queckenbekämpfung auf den Feldern ein annähernd äquimolares Gemisch von 3-AT mit Ammoniumthiocyanat (Amitrol T) empfohlen (Segal 1962), das auch bei Ackerwinde, Disteln, Bermudagras, Salzgras, wildem Roggen usw. gute Wirksamkeit besitzt. Gegen perennierende Unkräuter (z. B. Johnsongras) auch in Gewässern und an Ufern verwendet man erfolgreich ein Gemisch von 3-AT mit Dalapon (2,2-Dichloropropionsäure), gegen Nachwuchs im Getreide und Mais, zur Vernichtung des Unkrautes auf Bahndämmen und Gleisanlagen sowie gegen Quecken und Pfeilkresse im Weinbau die Kombination 3-AT mit Simazin (Domatol) und zum Lichten von Unterholz eine Mischung von 3-AT mit 2,3,6-TBA. Gegen Unkraut auf Wegen und Plätzen werden Gemische von 3-AT mit 2,4-D und TCA (Unkrautvernichter 2374) sowie mit 2,4-D und Atrazin (Vorox *i*) empfohlen. Im Gemisch mit Natriumchlorat 1 : 25 dient es zum Austrocknen der Mycele der Peronospora bei Tabakpflanzen (Zanardi 1962). Speziell gegen Winden im Weinbau wird die Kombination mit Simazin und MCPA (Domatol spezial *W*) als sogenanntes Nachlaufmittel angeboten (Kadisch 1965). Weitere Präparate bestehen aus Gemischen des 3-AT mit 2,4,5-T oder Monuron [3-(p-chlorophenyl)-1,1-dimethylharnstoff].

XI. Anwendungsweise

Damit das Erntegut keine 3-AT-rückstände enthält, soll die Anwendung seiner Präparate stets vorsichtig und unter genauer Beachtung der jeweiligen Gebrauchsanleitung erfolgen. So darf es u. a. in Preiselbeerkulturen nur innerhalb sieben bis zehn Tage nach der Ernte (Flemming 1959), in Baumbeständen, z. B. Apfel- und Birnenplantagen, nur vor dem Fruchtansatz oder nach dem Pflücken des Obstes und auf Mais- und Sojabohnenfeldern nach der Ernte sowie im Hackfruchtbau vor dem Einsäen angewandt werden (A.C.P. 1956). Im Gartenbau empfiehlt sich in den meisten Fällen eine 3-AT-behandlung mindestens zwei Wochen vor dem Säen oder Einpflanzen, während im Weinbau die Unkrautbekämpfung mit 3-AT-haltigen Mitteln frühestens erfolgen soll, wenn die Beeren Schrotkorngröße erreicht haben (Ende Juli, etwa drei Wochen nach der Blüte) (Eisenbarth 1965, Kadisch 1965).

XII. Analytik

Die z. Z. bekannten Nachweis- und Bestimmungsmethoden gründen sich vor allem auf der leichten Diazotierbarkeit des 3-AT und der Kupplung der Diazoverbindung zu Farbstoffen. Daneben wird noch seine mit Nitrosopentacyanoferraten auftretende Grünfärbung und seine Reduktionswirkung gegen Silbersalze analytisch ausgewertet.

a) Nachweismethoden

Zum schnellen und zuverlässigen Nachweis des 3-AT in Pflanzen- und Organextrakten, Böden usw. haben sich die papierchromatographischen Verfahren gut bewährt. GREEN und FEINSTEIN (1957) verwenden ein Gemisch von Isopropanol/Wasser/Ammoniak (28prozentig) = 80/10/10 als Fließmittel, entwickeln aufsteigend und besprühen dann das getrocknete Chromatogramm unter jedesmaligem Zwischentrocknen mit 0,3*N*-HCl, 0,5*M*-$NaNO_2$-lösung und einer 0,0025*M* Lösung von Chromotropsäure, wobei rotbraune Flecke erhalten werden. Mit diesem Verfahren sind noch etwa 0,2 μg. 3-AT nachweisbar.

ALDRICH und MCLANE (1957) entwickeln das Chromatogramm aufsteigend auf Whatman No. 1 Papier mit dem Fließmittel Äthanol/Wasser/*n*-Butanol = 1/1/4. Besprüht wird mit fünfprozentiger KNO_2-lösung und anschließend noch feucht mit einer gesättigten Lösung von Phenol in 20-prozentiger HCl. Dabei wird noch ~ ein μg. als gelber Fleck nachgewiesen. Die Methode ist auch für den 3-AT-nachweis in Milch geeignet.

Ebenfalls aufsteigend entwickelt RACUSEN (1958) seine Chromatogramme auf Whatman No. 1 Papier und gibt dafür als geeignete Fließmittel die Gemische an:

Methanol/Ameisensäure/Wasser = 80/15/5	(R_f = 0,60)
Isopropanol/Ammoniak/Wasser = 6/2/2	(R_f = 0,75)
80prozentiges Phenol	(R_f = 0,78)
n-Butanol/Essigsäure/Wasser = 27/7/17	(R_f = 0,55)
80prozentiges Pyridin	(R_f = 0,72)

Das getrocknete Chromatogramm wird mit einer einprozentigen $NaNO_2$-lösung in 80prozentigem Aceton, einer 20prozentigen acetonischen HCl, und nach zehn Minuten mit einer 0,25prozentigen Lösung des Dinatriumsalzes der *H*-säure (8-Amino-1-naphthol-3,6-disulfonsäure) in 50prozentigem Aceton besprüht, wobei der 3-AT-fleck kräftig rot auf weißem Untergrund erscheint. Mit dieser Methode sind auch die beiden Umwandlungsprodukte *X* und *Y* im Extrakt 3-AT-behandelter Blätter nachweisbar.

Auch MITCHELL (1960) beschreibt einen papierchromatographischen 3-AT-nachweis mit einem Gemisch von *p*-Dioxan/Wasser = 80/20 als Fließmittel. Nach etwa dreistündiger aufsteigender Entwicklung wird das Chromatogramm getrocknet und mit einer ammoniakalischen Lösung von Silbernitrat allein oder durch Nachsprühen mit Pyrogallollösung und anschließendes 15 Minuten langes Erhitzen auf 130° C. die Flecke sichtbar gemacht. Es können so noch etwa 0,5 μg. 3-AT ermittelt werden.

b) Bestimmungsmethoden

1. Titrimetrisch. — Nach dieser, speziell zur Bestimmung des 3-AT-gehaltes von Zubereitungen geeigneten, von SUTHERLAND (1964) mitgeteilten Methode wird das 3-AT mit *N,N*-Dimethylformamid extrahiert, der Extrakt mit überschüssiger 0,5 *N* HCl versetzt und dann mit Hilfe eines pH-Meters der nicht vom 3-AT gebundene Säureanteil zurücktitriert. Dabei werden die zum Erreichen der ersten Neutralitätsstufe bei pH 2,5 bis 2,9 und der zweiten Stufe bei pH 7,5 jeweils verbrauchten Mengen an 0,5 *N* NaOH getrennt abgelesen und aus diesen Werten anhand einer mit reinem 3-AT erstellten Eichkurve der Wirkstoffgehalt des Präparates errechnet.

2. Kolorimetrisch. — Diese Methoden zählen zu den empfindlichsten und sind deshalb auch für eine Restmengenbestimmung des 3-AT besonders geeignet. Je nach Art des Ausgangsmaterials und der erforderlichen Empfindlichkeit wurden mehrere verschiedene Arbeitsgänge beschrieben.

α) *In Böden.* — Nach SUND (1956) werden 100 g. der Bodenprobe mit 100 bis 250 ml. Wasser 15 Minuten lang geschüttelt und nach dem Filtrieren 25 oder 50 ml. des Extraktes in einem 100 ml.-Meßkolben auf etwa 70 ml. verdünnt, drei Tropfen zehnprozentige Natronlauge und zehn ml. Reagenslösung (siehe unten) zugegeben und mit Wasser zur Marke aufgefüllt. Nach 45 Minuten Stehen bei Raumtemperatur filtriert man und mißt die Extinktion des Filtrates genau eine Stunde nach Zugabe des Reagenses in einer Zwei-cm.-Küvette bei 650 nm. Die Auswertung erfolgt an Hand einer gleichartig mit Mengen von 0,1 bis 1,5 mg. 3-AT erstellten Eichkurve. Reagens:

59,6 g. Natriumnitrosopentacyanoferrat(II)-dihydrat zu 1 L. lösen
84,4 g. Kaliumhexacyanoferrat(II)-trihydrat zu 1 L. lösen
zehnprozentige Lösung von Natriumhydroxid
dreiprozentige Wasserstoffperoxidlösung

Die Lösungen 1. bis 4. werden im Volumenverhältnis 2 : 2 : 1 : 5 15 Minuten vor Gebrauch des Reagenses zusammengegeben und mit etwa 1,2 ml. Essigsäure auf 100 ml. Gemisch zum Einstellen des pH-Wertes auf $\sim$ 11, dem Optimum der Farbentwicklung, versetzt. Das Absorptionsmaximum des mit dem Reagens erhaltenen grünen Farbstoffes liegt bei 634 nm.

Diese in der Literatur zuerst beschriebene Bestimmungsmethode eignet sich zwar für reine oder mit nur geringen Verunreinigungen belastete 3-AT-lösungen, versagt jedoch in Gegenwart pflanzlicher Inhaltsstoffe, die reduzierend auf das Reagens wirken. In diesem Fall führen die auf Diazotierung und Farbstoffkupplung basierenden Verfahren besser zum Ziel.

β) *In Pflanzenteilen.* — Diese Arbeitsvorschriften unterscheiden sich vor allem durch die Art der Probenaufbereitung, die zur Kupplung verwendeten Substanzen und die damit zusammenhängende Empfindlichkeit und Spezifität der Farbstoffbildung.

$\alpha\alpha$) Die erste dieser Bestimmungsmethoden wurde von GREEN und FEINSTEIN (1957) entwickelt und verwendet die Chromotropsäure (4,5-Dihydroxynaphthalin-2,7-disulfonsäure) als Kupplungsreagens, wobei ein braunoranger Farbstoff erhalten wird, dem die Autoren die Formel

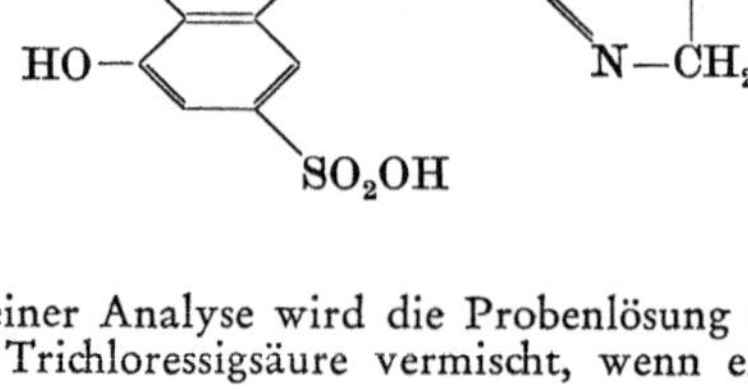

zuschreiben.

Zur Ausführung einer Analyse wird die Probenlösung mit dem gleichen Volumen zehnprozentiger Trichloressigsäure vermischt, wenn erforderlich filtriert, mit fünfprozentiger Trichloressigsäure verdünnt und fünf ml. der Verdünnung mit ein ml. 0,01 *M* $NaNO_2$-lösung versetzt. Unter Schütteln fügt man dann einen ml. 0,0025 *M* Chromotropsäurelösung zu, stellt 2,5 Minuten in kochendes Wasser, kühlt anschließend im Eisbad, und mißt die Extinktion der Farblösung bei 525 nm gegen einen Blindansatz. Erfassungsgrenze: zehn µg. 3-AT.

Obwohl durch die Trichloressigsäure viele störende Substanzen wie Proteine, Pektine usw. ausgefällt werden, beeinflussen doch noch einige Inhaltsstoffe der Probe die Farbentwicklung merklich. Ein weiterer Nachteil der Methode liegt darin, daß die Eichkurve nicht dem Lambert-Beer'schen Gesetz gehorcht, sondern parabelförmig verläuft, wobei sie sich mit steigenden 3-AT-mengen immer mehr einer Horizontalen nähert.

ββ) Den gleichen Mangel weist auch die von WINKLER (1959) für Preiselbeeren modifizierte Methode auf.

Das 3-AT wird zunächst aus einem mit verd. H_2O_2 bei Siedehitze oxydierten Methanolextrakt der Probe an einer sauren Amberlitesäule absorbiert. Das dann daraus mit Ammoniak gewonnene Eluat wird nach dem Einengen mit Trichloressigsäure auf pH zwei gebracht, auf 30 ml. eingestellt und über eine Aluminiumoxidsäule gereinigt. Mit der so gewonnenen Lösung wird wie bei *αα*) beschrieben die Farbentwicklung vorgenommen. Die Messung erfolgt bei 500 nm gegen einen Blindwert. Zur Auswertung wird eine Eichkurve mit ein bis 50µg. 3-AT aufgestellt.

γγ) HERRETT und LINK (1961) schlagen die 8-Amino-1-naphthol-3,6-disulfonsäure (*H*-säure) als Kupplungsreagens für das diazotierte 3-AT vor.

Die zu prüfenden Pflanzenteile werden im Mixgerät mit Wasser, Methanol oder wäßrigem Äthanol zerkleinert und der Brei zur Extraktion etwa zwei Stunden stehengelassen. Nach dem Filtrieren wird das Filtrat durch eine stark saure Amberlitesäule gegeben, mit Wasser nachgewaschen und dann das 3-AT mit Ammoniak eluiert. Nach dem Einengen wird das Eluat erneut über ein schwach saures Austauscherharz gegeben und dann daraus das 3-AT mit verdünnter Salzsäure eluiert. Drei ml. dieses Eluates werden im Eisbad mit 0,5 ml. einprozentiger $NaNO_2$-lösung unter Zusatz von 0,5 ml. 20prozentiger HCl diazotiert und dann 0,2 ml. 0,25prozentige *H*-säurelösung in 50prozentigem Aceton zugegeben. Nach 30 Minuten wird der entwickelte rote Farbstoff bei 538 nm gegen einen Blindansatz photometriert. Die Farbe bleibt bis zu etwa drei Minuten nach dem Entfernen aus dem Eis beständig.

δδ) Auch das *N*-(1-Naphthyl)-äthylendiamin ist als Kupplungskomponente für das diazotierte 3-AT geeignet. STORHERR und BURKE (1961) modifizierten die *U.S. Food and Drug Administration*-Preiselbeermethode, um sie auch für andere Früchte anwendbar zu machen.

Dazu zerkleinert man die Probe (rd. 200 g.) unter Zusatz von 300 ml. 95-prozentigem Äthanol und 20 g. Celite 545, filtriert, gibt zehn ml. 30prozentiges H_2O_2 zum Filtrat und erhitzt es 30 Minuten lang auf dem Dampfbad. Nach dem Abkühlen wird 30 Minuten lang mit 40 g. Amberlite 120 (*H*) durchgeschüttelt und durch ein Glaswollefilter dekantiert. Das Harz wird dann mehrmals mit 50 ml. Wasser unter Dekantieren ausgewaschen, das Glaswollefilter zum Harz in den Kolben zurückgegeben und 50 ml. Wasser und 20 ml. konzentriertes Ammoniak zugesetzt. Nach 30 Minuten Erhitzen auf dem Dampfbad wird die Flüssigkeit über ein Glaswollefilter dekantiert, das Harz mehrmals mit je 25 ml. 2 *N* Ammoniak ausgewaschen und das Filtrat auf etwa zehn ml. eingeengt. Zu dem abgekühlten Konzentrat werden 50 ml. Acetonitril und unter Rühren vier bis fünf g. Celite 545 gegeben. Dann wird durch ein Faltenfilter filtriert und dreimal mit je 50 ml. Acetonitril, das zehn Prozent Wasser enthält, nachgewaschen. Filtrat und Waschlösung werden im Luftstrom auf 20 ml. eingeengt, zehn bis 20 ml. Wasser zugefügt und erneut auf 20 ml. abgedunstet. Dies wird solange wiederholt, bis kein Acetonitril-Geruch mehr wahrnehmbar ist. Anschließend wird gekühlt, auf 20 ml. verdünnt, drei ml. H_2SO_4 (3+1) zugegeben und zehn Minuten lang gekocht, wobei das Volumen der Lösung um etwa zehn ml. abnimmt. Nach dem Abkühlen werden 50 ml. Wasser und ein g. Aktivkohle (Nuchar-*C* 190 *N*) zugesetzt und zehn Minuten lang kräftig gekocht. Die heiße Flüssigkeit wird filtriert und Glas und Filter mit 200 ml. heißer 0,75prozentiger H_2SO_4 gefolgt von 50 ml. heißem Wasser nachgewaschen. Die Filtrate werden auf etwa zehn ml. eingeengt und wenn noch gelb gefärbt erneut mit Aktivkohle aufgekocht und filtriert. Von dem genau aufgefüllten Filtrat werden fünf ml. zur Farbentwicklung entnommen, in einem 25 ml.-Meßkolben mit einem ml. Wasser und drei ml. H_2SO_4 (3+1) vermischt, 0,5 ml. 0,5prozentige $NaNO_2$-lösung zugegeben und zehn Minuten stehengelassen. Danach setzt man 0,5 ml. fünfprozentige Sulfaminsäurelösung und 0,5 ml. einprozentige Lösung von *N*-(1-naphthyl)-äthylendiamin-hydrochlorid zu, mischt gut durch, läßt fünf Minuten stehen und mißt gegen einen Blindansatz bei 455 nm. Erfassungsgrenze 0,025 p.p.m. 3-AT. Äthanol und Essigsäureäthylester stören die Farbentwicklung stark.

Eine noch vollständigere Abtrennung des 3-AT von störenden Begleitstoffen erreichen Storherr und Onley (1962) durch Adsorption an einer Cellulosesäule mit nachfolgender selektiver Elution.

Trotz ihrer Spezifität und z. T. recht großen Empfindlichkeit befriedigen die beschriebenen Bestimmungsmethoden nach Untersuchungen von Kröller (1961) nicht ganz. So stören z. B. viele Inhaltsstoffe der Pflanzenteile die Abtrennung oder die Farbentwicklung des 3-AT, einzelne Arbeitsgänge sind unübersichtlich und schlecht reproduzierbar, da Verluste auftreten können oder sich durch Nebenreaktionen Fehler nicht ganz vermeiden lassen. Bei den Methoden *αα*) und *ββ*) verläuft beispielsweise die Farbstoffkupplung nur sehr schlecht reproduzierbar, so daß selbst mit reinem 3-AT keine zuverlässige Eichkurve erstellt werden kann. U. a. schwankt bei 3-AT-konzentrationen von 1,5 μg./ml. die Extinktion bei 500 nm (Spektralphotometer) zwischen 0,032 und 0,052 und bei 490 nm (Elko) zwischen 0,028 und 0,046. Daneben nimmt auch die Farbdichte mit der Zeit stetig ab, so daß z. B. eine sofort nach der Kupplung abgelesene Extinktion von 0,448 innerhalb von 15 Minuten auf 0,416 und nach 25 Minuten auf 0,377 absinkt. Dies ist auch der Grund für die nur etwa drei Minuten anhaltende Beständigkeit des mit *H*-säure nach Methode *γγ*) erhaltenen Farbstoffes. Die Ursache dieses „Ausbleichens" ist der beim Diazotieren nicht verbrauchte Überschuß an salpetriger Säure, der wegen zu langsam verlaufender Reaktion weder durch Harnstoff, Hydrazin, oder Hydroxylamin noch durch Sulfaminsäure völlig zerstört werden kann. Dies läßt sich dagegen schnell und quantitativ mit Ascorbinsäure erreichen, so daß es nunmehr gelingt, die Farbdichte der Kupplungsprodukte über 30 Minuten lang bei Raumtemperatur konstant zu halten. Ein weiterer Mangel der Methoden *αα*) und *ββ*), die parabelförmig verlaufenden Eichkurven, läßt sich durch die Wahl optimaler Reaktionsbedingungen (Temperatur und Zeit) ebenfalls beheben, wobei eine Reaktionszeit von 20 Minuten bei 50° C. zu einer linearen, dem Lambert-Beer'schen Gesetz gehorchenden Eichkurve führt. Selbst Spuren von Schwermetallen und geringe Reagensüberschüsse stören dann den Verlauf nicht mehr. Die Erfassungsgrenze der Methode *ββ*) läßt sich damit auf 0,5 μg. 3-AT/ml. verbessern.

Ein weiterer wichtiger Faktor für die Reproduzierbarkeit der Ergebnisse einiger Bestimmungsmethoden wurde in der Konstanthaltung des pH-wertes bei der Farbstoffbildung ermittelt. Insbesondere das mit *H*-säure nach Methode $\gamma\gamma$) erhältliche rote Kupplungsprodukt wechselt Farbton und -dichte mit dem pH der Lösung, wobei eine Verschiebung zum sauren Gebiet über gelbstichig rote zu orangen Färbungen und eine solche zum alkalischen Teil zu stark aufgehellten hellroten Farben führt. Das Maximum an Intensität, ein kräftiges violettstichiges Dunkelrot, wird bei pH 9 erhalten. Seine Absorptionskurve steigt ab 471 nm steil an, erreicht bei 525 nm ein spitzes Maximum und fällt dann wieder bis 569 nm steil ab. Die bei 525 nm gemessenen Extinktionen sind der 3-AT-menge proportional und gehorchen dem Lambert-Beer'schen Gesetz. Spuren von Eisen, Chrom, und Titan sowie geringe Reagensüberschüsse stören die Reaktion nicht, jedoch wird die Farbstoffbildung von Mangan geschwächt und von Kupfer verhindert, so daß streng auf Abwesenheit dieser Elemente in den Reagentien zu achten ist.

Ein besonders stark störender Pflanzen-Inhaltsstoff ist das Tannin, das schon in kleinsten Mengen die Farbstoffbildung des diazotierten 3-AT mit allen Kupplungskomponenten hemmt oder sogar völlig verhindert. Von 24 μg. 3-AT konnte so mit *H*-säure z. B. in Gegenwart von 50 μg. Tannin überhaupt nichts wiedergefunden werden, während bei fünf μg. Tannin rd. 17 μg. 3-AT, bei einem μg. Tannin 21 μg. 3-AT, bei 0,5 μg. Tannin 22,5 μg. 3-AT, und bei 0,1 μg. Tannin 23,8 μg. 3-AT bestimmt wurden. Auch die Ascorbinsäure und ihr Oxydationsprodukt stören durch Beeinflussung der Diazotierungsreaktion die Farbstoffbildung. Eine Abtrennung dieser Verbindungen z. B. aus Preiselbeersaft läßt sich weder durch Fällung mit Bleiessig, Barytwasser, Aluminium- oder Quecksilbersalzen noch durch Ausschütteln mit Äther, Chloroform, und Amylalkohol oder durch Adsorption an Austauscherharzen, Aluminiumoxid oder Kieselgel verschiedener Aktivitätsstufen bewerkstelligen, gelingt jedoch nach vorheriger Reduktion mit Natriumdithionit. Eine besondere Bedeutung für die Empfindlichkeit der Farbreaktion kommt schließlich noch der Kupplungsfähigkeit der einzelnen Komponenten mit dem diazotierten 3-AT zu. Während z. B. Phenylhydrazinsulfonsäure, Sulfanilsäure, 1-Amino-2-naphthol-4-sulfonsäure, β-Naphthochinon-sulfonsäure, 2-Benzolazo-1,8-dihydroxynaphthalin-3,6-disulfonsäure, Äthyl-α-naphthylamin, α-Naphthylamin, *p*-Phenylendiamin, α-Naphthol, und *p*-Nitrobenzol-azo-α-Naphthol mit diazotiertem 3-AT nicht zu Farbstoffen zusammentreten, werden schwache Farbtöne erhalten mit β-Naphthol (gelborange), α-Naphthylaminsulfonsäure (rot), 7-Amino-naphthol-1-sulfonsäure-3 (hellorange), 2-Naphthol-6,8-disulfonsäure (hellgelb), 2,3-Dihydroxynaphthalin-6-sulfonsäure (orange) und Resorcin-4,6-disulfonsäure (hellgelb). Kräftige, und damit für eine kolorimetrische Bestimmung geeignete Farbstoffe werden dagegen mit den bereits unter α), β), γ) und δ) genannten Kupplungskomponenten sowie u. a. mit β-Naphthylamin (orange, Absorptionsmaximum 475 nm), *N*-Äthyl-*N*-benzyl-anilin-3-sulfonsäure (orange, 505 nm), 8-Amino-1-naphthol-5,7-disulfonsäure (erdbeerrot, 530 nm), 2-Naphthol-3,6-disulfonsäure (goldgelb, 460 nm), und Phenol (gelb, 460 nm) erhalten.

εε) Nach diesen Untersuchungen lautet die verbesserte Arbeitsvorschrift für die Restmengen-Bestimmung von 3-AT:

Geräte

Mixgerät
Abnutschvorrichtung mit Büchnertrichter ∅ 150 mm.
Saftfilter S. u. S. Nr. 572 ∅ 15 cm.
Meßzylinder 1500 ml.
Chromatographierröhren mit Hahn und Siebplatte ∅ 20 mm., 650 mm. lang und ∅ 20 mm., 250 mm. lang
Infrarot- oder Wasserbad
Thermostat
Stoppuhr
Spektralphotometer
Reagensgläser ca. 180 mm. lang, ∅ 20 mm.
Pipetten
Bechergläser

Chemikalien

Methanol p. A.
Methanol 80 Vol.prozentig
Doppelt destilliertes Wasser für alle Verdünnungen
Natriumdithionit p. A.
Ionenaustauscher stark sauer (z. B. „Merck“ Nr. 1)
Seesand p. A.
Glaswolle
2*N*-Ammoniak
Aluminiumoxid sauer zur Chromatographie (z. B. Woelm)
Eisen-III-chloridlösung einprozentig
1*N*-Salzsäure
5*N*-Salzsäure
0,01 *M*-Natriumnitritlösung
Chromotropsäurelösung, 100 mg. Na-salz/250 ml. oder eine 0,1prozentige Lösung einer der anderen Kupplungskomponenten (bei *H*-säure außerdem noch eine fünfprozentige Boraxlösung)
Ascorbinsäurelösung zweiprozentig
Aminotriazol-Standardlösung, 50 mg. 3-AT in 500 ml.

Vorbereitung der Probe

500 g. Preiselbeeren (oder andere Früchte oder Pflanzenteile) werden unter Zugabe von 200 ml. Methanol im Mixgerät gut zerkleinert und der erhaltene Brei mit etwa 500 ml. 80prozentigem Methanol in ein 2000 ml.-Becherglas quantitativ überspült. Anschließend läßt man zur Fällung von Pektinen und Schleimstoffen eine Stunde stehen und saugt dann den Extrakt durch ein Saftfilter ab. Der Filterrückstand wird zwei- bis dreimal mit zusammen etwa 200 ml. 80prozentigem Methanol gut durchmischt und jedesmal bis zur Trockne abgesaugt. Das Filtrat füllt man dann im Meßzylinder mit 80prozentigem Methanol auf und bereitet zur Weiterverarbeitung dieses Extraktes eine Ionenaustauschersäule wie folgt vor:

Auf die Siebplatte eines 650 mm. langen Chromatographierrohres gibt man etwa 1/2 cm. Glaswolle und darauf etwa acht cm. hoch mit Wasser vorgequollenes, stark saures Ionenaustauscherharz. Darüber wird wieder Glaswolle und etwa ein cm. hoch Seesand geschichtet.

Zur Aktivierung des Austauschers werden mit stets gleichbleibender Geschwindigkeit von zehn ml/Minute 200 ml. 5*N*-Salzsäure, dann 150 ml. Wasser zum Nachwaschen, anschließend 200 ml. 2*N*-Ammoniak und wieder 150 ml. Wasser durch die Säule gegeben. Schließlich läßt man erneut 200 ml. 5*N*-Salzsäure und zum Nachspülen 150 ml. Wasser durchlaufen und spült kurz vor Gebrauch mit

100 ml. 80prozentigem Methanol durch. Dabei ist darauf zu achten, daß die Seesandschicht stets etwa 1/2 cm. hoch mit Flüssigkeit bedeckt bleibt und keine Luftblasen in die Austauschersäule geraten.

Danach werden 300 ml. des Extraktes im Becherglas mit zwei g. festem Natriumdithionit versetzt und bis zur völligen Lösung zum Sieden erhitzt. Nach dem Abkühlen auf Raumtemperatur wird die Flüssigkeit durch die vorbereitete Austauschersäule gegeben (Durchlaufgeschwindigkeit zehn ml./Minute) und nacheinander mit 100 ml. 80prozentigem Methanol und 100 ml. Wasser nachgewaschen. Das Perkolat wird verworfen. Das am Harz festgehaltene 3-AT wird mit 100 ml. 2*N*-Ammoniak und anschließend mit 100 ml. Wasser eluiert und das Eluat im 250 ml.-Becherglas auf dem Wasserbad im Luftstrom auf etwa 30 ml. eingeengt, wobei sich das Ammoniak verflüchtigt. Das so erhaltene Konzentrat wird dann zum Entfernen störender Stoffe durch eine Aluminiumoxidsäule gegeben. Dazu wird auf die Siebplatte eines 250 mm. langen Rohres etwas Glaswolle und darauf eine etwa fünf cm. hohe Schicht von aktivem saurem Aluminiumoxid gebracht, das unter stetem Schütteln und leichtem Klopfen portionsweise eingefüllt wird. Dann wird unter geringem Überdruck Wasser durch die Säule gegeben, bis es völlig klar abläuft. Anschließend schickt man das konzentrierte Eluat durch die Säule und spült mit 50 ml. Wasser nach. Die so gereinigte Lösung, die möglichst farblos sein soll, wird mit einem ml. einprozentiger Eisen-III-chloridlösung versetzt, auf dem Infrarotbad mittels eines schwachen Luftstromes auf etwa zehn ml. eingeengt und nach dem Abkühlen mit vier bis fünf Tropfen 2*N*-Ammoniak versetzt. Das sich flockig abscheidende Eisenhydroxid filtriert man durch ein kleines Filter ab und fängt das Filtrat in einem 25 ml.-Meßkolben auf, wäscht mit wenig Wasser nach und füllt zur Marke auf.

Kolorimetrische Bestimmung mit Chromotropsäure

Von der so erhaltenen Analysenlösung pipettiert man zwei ml. in ein Reagensglas und setzt gleichzeitig mit zwei ml. Wasser eine Blindprobe sowie zwei Vergleichsproben von zehn und 20 μg. 3-AT in jeweils zwei ml. Wasser an. Zu jeder dieser Proben werden nacheinander je drei ml. 1*N*-Salzsäure und 0,2 ml. 0,01 *M*-Natriumnitritlösung unter jedesmaligem Schütteln gegeben. Dann wird in einem Wasserbad genau 15 Minuten lang auf 50° C. erwärmt, sofort in kaltem Wasser gekühlt und unter Schütteln je 1,5 ml. Chromotropsäurelösung zugesetzt. Unmittelbar darauf werden die Gläser genau 2 1/2 Minuten in kochendes Wasser gestellt, dann in Eiswasser auf Raumtemperatur gekühlt. Nach Zusatz von je 0,5 ml. Ascorbinsäurelösung werden die Proben in einer zwei cm.-Küvette bei 505 nm gegen die Blindprobe gemessen. Da die Extinktionen des Farbstoffes das Lambert-Beer'sche Gesetz erfüllen, kann der 3-AT-gehalt der Probe mit der mittels der Vergleichsproben erstellten Eichkurve ermittelt werden. Die Konzentration in p.p.m. ergibt sich dann durch Multiplikation des abgelesenen Wertes in μg. mit 0,125. Die Erfassungsgrenze für 3-AT in Preiselbeeren nach dieser Methode liegt bei 0,1 p.p.m. ± 10 Prozent.

Kolorimetrische Bestimmung mit H-säure

Weitaus empfindlicher als die vorbeschriebene Methode ist die auf der Kupplung des diazotierten 3-AT mit *H*-säure beruhende Bestimmung.

Dazu werden zwei ml. der nach obiger Vorschrift erhaltenen Analysenlösung mit 0,3 ml. 1 *N*-Salzsäure und 0,2 ml. 0,01 *M*-Natriumnitritlösung 15 Minuten lang auf 50° C. erwärmt. Nach dem Erkalten werden 0,4 ml. 1 *N*-Natronlauge, 1 ml. fünfprozentige Boraxlösung und anschließend 0,2 ml. 0,1prozentige Lösung des Natriumsalzes der *H*-säure (8-Amino-1-naphthol-3,6-disulfonsäure) zugegeben, umgeschüttelt und fünf Minuten stehengelassen. Nach Zugabe von 0,5 ml. zweiprozentiger Ascorbinsäurelösung wird nach zehn Minuten in einer zwei cm.-Küvette gegen den mit zwei ml. Wasser gleichartig angesetzten Blindwert bei 525 nm gemessen. Auch hierbei erfolgt die Auswertung an Hand einer gleichartig mit reinen 3-AT-lösungen erstellten Eichkurve. Die Erfassungsgrenze liegt bei 0,05 μg./ml. entsprechend 0,01 p.p.m. 3-AT.

Kolorimetrische Bestimmung mit β-Naphthylamin

Etwas weniger empfindlich, dafür aber auch weniger störanfällig ist die durch Farbstoffkupplung mit β-Naphthylamin mögliche Bestimmung. Man versetzt dazu zwei ml. der nach obiger Vorschrift bereiteten Analysenlösung mit 0,3 ml. 1*N*-Salzsäure und 0,2 ml. 0,01*M*-Natriumnitritlösung und erwärmt 15 Minuten lang auf 50° C. Anschließend wird innerhalb fünf Minuten auf Raumtemperatur abgekühlt und unter Umschütteln nacheinander 0,25 ml. 1*N*-Natronlauge, 0,2 ml. 0,1prozentiger β-Naphthylaminhydrochlorid-lösung und 0,2 ml. zweiprozentige Ascorbinsäurelösung zugegeben. Die Extinktion wird nach zehn Minuten in einer ein cm.-Küvette bei 480 nm gegen einen mit zwei ml. Wasser gleichartig angesetzten Blindwert gemessen. Ausgewertet wird mit einer entsprechend aufgestellten Eichkurve. Die Erfassungsgrenze liegt bei 0,2 µg./ml. entsprechend etwa 0,05 p.p.m.

Kolorimetrische Bestimmung mit N-Äthyl-N-benzyl-anilin-3-sulfonsäure

Als weitere Möglichkeit sei schließlich noch die folgende Vorschrift mitgeteilt, die etwa die gleiche Empfindlichkeit wie die vorhergehende besitzt.

Hierzu werden zwei ml. der nach obiger Vorschrift gewonnenen Lösung mit drei ml. 1*N*-Salzsäure und 0,2 ml. 0,01*M*-Natriumnitritlösung 15 Minuten lang auf 50° C. erwärmt. Nach dem Erkalten werden 0,5 ml. 0,1prozentige Lösung der *N*-Äthyl-*N*-benzylanilin-3-sulfonsäure zugegeben und wieder zehn Minuten auf 50° C. erwärmt. Nach dem Abkühlen gibt man 0,5 ml. zweiprozentige Ascorbinsäurelösung zu und mißt die Extinktion in einer ein-cm.-Küvette gegen den mit Wasser angesetzten, schwach hellgelben Blindwert bei 505 nm. Empfindlichkeit: 0,2 µg. 3-AT/ml. entsprechend etwa 0,05 p.p.m.

Nach diesen Arbeitsvorschriften kann der 3-AT-gehalt frischer und eingedoster Preiselbeeren sowie aller Obstarten und Pflanzenteile im Empfindlichkeitsbereich von 0,1 bis hinab zu 0,01 p.p.m. unter wahlweisem Einsatz der jeweils geeignetsten Farbreaktion im sauren oder alkalischen Medium mit einer durchschnittlichen Fehlerbreite von ± zehn Prozent bestimmt werden.

c) Biologischer Test

Ein auf der Wachstumshemmung beruhender Test mit Soja- und Sonnenblumensamen, der es gestattet, verschiedene Herbicide nebeneinander und ihre Restmengen im Boden usw. auf Grund ihres spezifischen Einflusses auf die Keime zu erkennen, wird von Gowing (1959) beschrieben.

XIII. Toleranzen und Rückstände

Wie bereits im Abschnitt über die Toxikologie erläutert, ist für 3-AT-Rückstände in allen Lebensmitteln sowohl in U.S.A. als auch in den meisten europäischen, vor allem den EWG-Ländern, die Toleranz auf null festgesetzt worden. Das heißt in die Praxis übertragen, daß mit der z. Z. empfindlichsten Bestimmungsmethode, die gleichzeitig möglichst selektiv und zuverlässig sein soll, kein 3-AT in der Probe mehr erfaßbar sein darf. Nach dem augenblicklichen Stand der Analytik liegt diese Grenze bei 0,01 p.p.m. Unberücksichtigt bleiben hierbei jedoch — wie es auch bei den meisten anderen Wirkstoffen der Fall ist — viele Umwandlungsprodukte (Metabolite) des 3-AT, die nicht mehr die charakteristischen Reaktionen der freien Base zeigen und deshalb einer quantitativen Bestimmung entgehen. Da diese, in ihrer chemischen Struktur meist noch nicht aufgeklärten Komplexe im Experiment häufig gegenüber dem unveränderten 3-AT eine stärkere toxische Wirksamkeit sowohl bei Pflanzen als auch vor allem bei Tieren und damit wohl auch beim Menschen entfalten, ist eine gesundheitliche Beurteilung ihrer im Lebensmittel verbleibenden Restmengen äußerst schwierig. Neben dem Problem einer umfassenden toxikologischen Überprüfung und Beurteilung der Metabolite tritt die Schwierigkeit ihrer analytischen Erfassung im Rahmen der Lebensmittelüberwachung besonders hervor. Aufgaben, die bis jetzt in befriedigender Weise noch nicht gelöst sind. Im Hinblick hierauf muß deshalb um so mehr gefordert werden, daß bei Anwendung des 3-AT alle Vorschriften bezüglich Art, Menge, und vor allem Zeit der Anwendung besonders streng beachtet werden. Bei Einhaltung derselben fanden Ercegovich und Frear (1964) keine Rückstände in Mais, Bohnen, Spinat, und Rettichen. Auch in Äpfeln und Birnen wurden bei richtiger Anwendung des 3-AT keine Restmengen gefunden. Anders dagegen liegen die Verhältnisse bei Preiselbeeren. Nach den sehr gründlichen Untersuchungen von Onley *et al.* (1963) in Preiselbeersümpfen von Oregon fanden sich sowohl beim Einsatz der empfohlenen Menge von acht lb./Acre als auch bei der Hälfte davon unabhängig vom Zeitpunkt der Anwendung stets beachtliche Restmengen 3-AT in den geernteten Beeren. Wurde das 3-AT im Frühjahr (Mai) angewendet, so enthielten die in der Zeit vom August bis Oktober geernteten Beeren zwischen 0,2 und 0,6 p.p.m. Restmenge, bei Einsatz der halben Wirkstoffmenge zwischen 0,1 und 0,3 p.p.m. Bei Spritzungen im Oktober nach der Ernte wurde im folgenden Jahr immer noch 3-AT in Mengen von 0,3 bis 0,4 p.p.m. im Erntegut ermittelt, bei Spritzung mit der halben Menge etwa 0,2 p.p.m.

Untersuchungen des Bodens, der Preiselbeerpflanze und ihrer Wurzeln ergaben folgendes Bild: Sofort nach der Anwendung im Oktober beginnt der 3-AT-gehalt des Bodens stetig abzunehmen und erreicht nach etwa 12 Monaten Werte von etwa 0,05 p.p.m. Die ebenfalls in der Pflanze zunächst sehr hohen 3-AT-mengen von bis zu rd. 182 p.p.m. nehmen sehr schnell ab, bleiben dann aber ungefähr acht Monate lang ziemlich konstant bei etwa sieben p.p.m., um dann zum Sommer auf zwei bis drei p.p.m. abzunehmen. Auch in der Wurzel läßt sich 3-AT bis zu einem p.p.m nachweisen. Diese Ergebnisse zeigen also, daß das 3-AT bis zu 12 Monaten nach seiner Anwendung im Boden und in sämtlichen Teilen der Preiselbeerpflanze in signifikanten Mengen enthalten ist, wobei es von Strauch und Wurzel in viel höheren Konzentrationen gespeichert wird als vom Boden. Das gespeicherte 3-AT geht dann in die Beeren über, in denen es ebenfalls noch 12 Monate nach der letzten Anwendung nachgewiesen werden kann. Auch ein Herabsetzen der verwendeten 3-AT-menge ändert wenig an dieser Tatsache, so daß entsprechend der Toleranzgesetzgebung alle innerhalb eines Jahres nach einer 3-AT-anwendung geernteten Preiselbeeren nicht verkehrsfähig sind. Da nach Miller und Hall (1961) das 3-AT — wahrscheinlich in Form eines wasserlöslichen Metaboliten — auch in die Samen der Baumwollpflanze eindringt und sich dort in Mengen bis zu über 50 p.p.m. ablagert, muß damit gerechnet werden, daß auch die Samen anderer 3-AT-behandelter Pflanzen diesen Stoff enthalten. Orientierende Versuche mit Most aus 3-ATgespritzten Weinpflanzungen erbrachten dagegen keinen Anhaltspunkt für nachweisbare Restmengen.

Zusammenfassung

Über die phytotoxischen Eigenschaften des bereits 1888 synthetisierten 3-Amino-1,2,4-triazols (3-AT) wurde erstmals 1955 ausführlich berichtet. Wegen seiner ausgezeichneten herbiciden Wirksamkeit findet es in großem Umfang im Obst-, Wein-, Acker- und Gartenbau sowie zur Unkrautvertilgung auf Straßen, Plätzen, Bahndämmen, in Teichen und Wasserläufen Verwendung. Ein besonderes Anwendungsgebiet sind die Preiselbeersümpfe in U.S.A. und die Baumwollplantagen, wo es als Entblätterungsmittel zur Erleichterung der Ernte eingesetzt wird. Durch Hemmen oder Blockieren der meisten Enzymsysteme und Störung der Chlorophyllbildung über Eingriffe in die Chloroplastenentwicklung führt das leicht auf dem wäßrigen Weg in die Zellen des Phloems und Xylems eindringende und frei in der Pflanze mit den Säfteströmen bewegliche 3-AT schnell zur Chlorose und zum Absterben. Durch Speicherung in den Wurzeln, Knollen, und Triebspitzen verbleibt es über lange Zeiträume in den Pflanzen, während es im Boden je nach Beschaffenheit in mehr oder weniger kurzen Zeiträumen von den Mikroorganismen abgebaut und von den Inhaltsstoffen gebunden wird. Auf Grund seiner großen Reaktionsfähigkeit bildet es in der lebenden Zelle mit den verschiedensten Stoffklassen eine große Zahl von Umwandlungsprodukten (Metaboliten) mit z. T. erheblich gesteigerter biologischer Wirksamkeit. Seine akute orale Toxizität gegen Warmblütler ist mi LD_{50} 14,7 g. pro kg. bei Mäusen und 25 g./kg. bei Ratten gering, doch vermag es im chronischen Fütterungsversuch mit hohen Dosen eine goitrogene Wirkung zu entfalten, wobei die stärksten Veränderungen der Schilddrüse als Adenocarcinom interpretiert werden. Aus diesem Grund dürfen die Lebensmittel keine nachweisbaren 3-AT-Restmengen enthalten, was besonders zu Beginn seiner Anwendung zu großen Beschlagnahmeaktionen bei Preiselbeeren und zu ihrer Vernichtung führte. Um nach der Behandlung Restmengen des 3-AT und damit Verluste der Ernte auszuschließen, muß die Anwendung der Präparate genau nach Dosierungsvorschrift und nur zu den dafür angegebenen Jahreszeiten erfolgen. Im allgemeinen ist dies der Herbst sofort nach der Ernte, in einigen Fällen auch das Frühjahr vor dem Einsäen oder Einpflanzen. Die Handelspräparate, die meist 50 Prozent 3-AT enthalten und unter den verschiedensten Namen angeboten werden, gestatten einen Einsatz als Sprüh- oder Gießmittel, als Stäubemittel, oder als Zusatz zu Gewässern. Da trotz des breiten Wirkungsspektrums 3-AT bei einigen Unkrautarten nicht zum Erfolg führt, wird es, auch um gleichzeitig seine Aufwandmenge herabzusetzen, mit anderen Herbiciden kombiniert. Analytisch wird die heterocyclische, in zwei desmotropen Formen vorkommende Base, die leicht in Wasser und Alkoholen löslich ist, mit Säuren Salze und mit vielen zwei- und dreiwertigen Metallen sehr stabile chelatartige Komplexe bildet, entweder auf papierchromatographischem Wege nachgewiesen oder kolorimetrisch — vorwiegend durch Diazotieren und Kuppeln mit verschiedenen Komponenten zu charakteristischen Farbstoffen — quantitativ bestimmt. Die Erfassungsgrenze der zur Zeit empfindlichsten Bestimmungsmethode liegt bei 0,01 p.p.m. Während in den meisten einjährigen Nutzpflanzen in Obstarten und Wein bei richtiger Anwendungszeit und Dosierung keine Rückstände festgestellt werden, können Preiselbeeren, selbst noch ein Jahr nach der Anwendung, durch Speicherung in allen Pflanzenteilen signifikante 3-AT-Restmengen enhalten, so daß sie nach den Restmengen-Bestimmungen der meisten Länder nicht verkehrsfähig sind. Da die meisten Analysenverfahren nur das freie, nicht umgewandelte 3-AT erfassen, ist damit zu rechnen, daß auch bei scheinbarer Abwesenheit von Resten erhebliche Mengen von Metaboliten mit z. T. noch stärker toxischer Wirkung als das 3-AT in den behandelten Pflanzen, ihren Früchten und Samen anwesend sind.

Summary*

The phytotoxic properties of 3-amino-1,2,4-triazol (3-AT) that was synthetized in 1888 were first described in 1955. Owing to its notable herbicidal action, it finds very wide use in orchards, vineyards, agriculture and horticulture, as weedkiller for streets, squares, platforms, and in ponds and streams. A special application is in cranberry fields in the U.S.A. and the cotton plantations where it is used as a defoliant for making the harvesting easier. The moving 3-AT easily penetrates by means of the aqueous phase in the phloem and xylem cells and, after being released in the plant by the sap streams, rapidly causes chlorosis and decay due to inhibition or blocking of the main enzymes and disturbing the chlorophyll formation by interference with the development of the chloroplasts. While it persists during long periods in the plants due to storage in the roots, the bulbs, and the sprouts, it is destroyed in the soil by microorganisms and absorbed within a more or less short lapse of time, according to the soil composition. Owing to its strong reactivity, it forms in the living cell and with a great many different classes of compounds a great number of transformation compounds (metabolites) which have a considerably enhanced biological activity. Its acute oral toxicity for mammals is low: 14.7 g./kg. for the mouse and 25 g./kg. for the rat; however, at high dosages in chronic ingestion tests it may involve a goitrogenic action, the most important alterations of the thyroid glands being interpreted as adenocarcinomas. Therefore, no detectable residue of 3-AT is allowed to be present in foodstuffs; this has resulted, particularly at the beginning of its use, in important seizures of cranberries and their subsequent destruction. In order to avoid 3-AT residues following the treatment and the consequent loss of the crop, the compound must be used scrupulously according to the directions and only during certain definite periods. Generally, the most suitable period is the one in the fall immediately following the harvesting; however, in some cases, spring prior to sowing or planting may be suitable. The marmercial formulations, which generally contain 50% of 3-AT and are preketed under different names, can be used by spraying, sprinkling, or

* Translated by S. Dormal-van den Bruel

dusting, or added to water. Although 3-AT has a wide range of activity, it is not efficacious on some kinds of weeds; therefore, it may be used in admixture with other weedkillers in order to reduce its application dosage. From the analytical standpoint, 3-AT is an heterocyclic base with two desmotropic forms, readily soluble in water and alcohols, and reacting with the acidic salts and many bi- and trivalent metals to form chelation complexes. It is quantitatively determined either by paper chromatography or by colorimetry — mainly by diazotation and coupling with various compounds to form distinctive dyes. The limit of detectability of the most sensitive existing method is 0.01 p.p.m. While most of the annual cultivated plants and fruits and wines do not contain any detectable residue, when the period of application and the dosages are correctly chosen, cranberries, on the other hand, still may show, one year after the application, significant amounts of 3-AT residues in the different parts of the plant; this means that, according to the residue determinations performed in most countries, this fruit is not commercialized. Since most of the methods of analysis only detect the free 3-AT and not its metabolites, one may assume that, even in the absence of apparent residues, significant amounts of metabolites partially endowed with toxic properties even greater than those of 3-AT may be present in the treated plants, the fruits, and the seeds.

Résumé *

On a rapporté en détail et, pour la première fois, en 1955 les propriétés phytotoxiques du 3-amino-1,2,4-triazole (3-AT) synthétisé en 1888. En raison de son activité herbicide remarquable, il trouve un usage très étendu dans les vergers, les vignobles, en agriculture et horticulture tout comme pour le désherbage des rues, places et quais de gares, ainsi que dans les étangs et les cours d'eau. Les cultures d'airelles, aux Etats-Unis, et les plantations de coton constituent un domaine d'application particulier où il est utilisé comme défoliant pour faciliter la récolte. Le 3-AT étant mobile, pénétrant facilement par voie aqueuse dans les cellules du phloème et du xylème et étant libéré dans la plante avec les courants de sève, entraîne rapidement la chlorose et la mort par inhibition ou blocage des principaux systèmes enzymatiques et perturbation de la formation de la chlorophylle par interférence dans le développement des chloroplastes. Il persiste durant de longues périodes dans les plantes par accumulation dans les racines, les bulbes et les pousses, tandis que, dans le sol, il est détruit par les microorganismes et absorbé par les substances qu'il contient endéans un laps de temps plus ou moins court, selon la constitution du sol. En raison de son grand pouvoir réactionnel, il forme dans la cellule vivante, avec les catégories les plus diverses de produits, un grand nombre de produits de transformation (métabolites) qui ont une activité biologique considérablement accrue. Sa toxicité aiguë par voie orale pour les animaux à sang chaud est faible: 14,7 g/kg pour la souris et 25 g/kg pour le rat; cependant, il peut exercer une action sur la genèse du goître lors des essais d'ingestion chroni-

* Traduit par S. DORMAL-VAN DEN BRUEL

que à fortes doses, où les plus importantes modifications des glandes thyroïdes ont été interprétées comme étant des adéno-carcinomes. Pour cette raison, les aliments ne peuvent contenir aucun résidu décelable de 3-AT, ce qui a entraîné, en particulier au début de son utilisation, d'importantes confiscations pour les airelles et la destruction de leurs stocks. Pour éviter les résidus de 3-AT à la suite du traitement et, par conséquent, la perte de la récolte, l'application de la préparation doit se faire exactement selon la prescription des doses et seulement à des époques déterminées de l'année. En général, l'époque requise est l'automne, immédiatement après la récolte; dans quelques cas, c'est le printemps avant le semis ou la plantation. Les préparations commerciales, qui contiennent le plus souvent 50% de 3-AT et qui sont présentées sous différents noms, peuvent être utilisées en pulvérisation ou arrosage, en poudrage ou être ajoutées aux eaux. Etant donné que, malgré son large spectre d'action, le 3-AT n'est pas efficace sur quelques espèces de mauvaises herbes, il est aussi combiné à d'autres herbicides, ce qui permet de réduire sa dose d'emploi. Du point de vue analytique, le 3-AT est une base hétérocyclique qui se présente sous deux formes desmotropes, est aisément soluble dans l'eau et les alcools, et forme des complexes de chélation très stables avec les sels acides et avec beaucoup de métaux bi- et trivalents. Il est déterminé quantitativement soit par chromatographie sur papier, soit par colorimétrie — surtout par diazotation et copulation avec divers composés pour former des colorants caractéristiques —. La limite de détectabilité de la méthode actuellement la plus sensible est de 0,01 ppm. Alors qu'on ne décèle aucun résidu dans la plupart des plantes cultivées annuelles, dans les fruits et le vin, si le moment de l'application et les doses d'emploi sont corrects, les airelles, par contre, peuvent encore contenir un an après le traitement des résidus non négligeables de 3-AT dans toutes les parties de la plante, de telle sorte que, d'après les déterminations des résidus faites dans la plupart des pays, ces fruits ne peuvent être commercialisés. Comme la plupart des méthodes d'analyse mettent seulement en évidence le 3-AT libre et non métabolisé, il faut présumer que, même en l'absence de résidus apparents, des quantités importantes de métabolites doués partiellement d'une action toxique encore plus forte que celle du 3-AT sont présents dans les plantes traitées, les fruits et les semences.

Literatur

Aaronson, S.: Mode of action of 3-amino-1,2,4-triazole on photosynthetic microorganisms. J. Protozool. **7**, 289 (1960).

A.C.C. *(American Cyanamid Company):* Aminotriazole. Tech. Data Sheet, Feb. 1955.

A.C.P. *(American Chemical Paint Company):* Soil residual after amizol application. Tech. Service Data Sheets H 56 (1955), H 64 (1956), H 68 (1957).

Aldrich, F. D.: Some indications of antagonism between 3-amino-1,2,4-triazole and purine and pyrimidine bases. Abstr. 1958 Meeting Weed Soc. Amer. p. 32 (1958).

—, and S. R. McLane, Jr.: A paper chromatographic method for the detection of 3-amino-1,2,4-triazole in plant tissues. Plant Physiol. **32**, 153 (1957).

Aldrich, R. J.: A preliminary evaluation of weed control in cranberries. Proc. Amer. Cranberry Growers Assoc. **86**, 7 (1956).

ALEXANDER, N. M.: Iodide peroxidase in rat thyroid and salivary glands and its inhibition by antithyroid compounds. J. Biol. Chem. **234**, 1530 (1959a).
— Antithyroid action of 3-amino-1,2,4-triazole. J. Biol. Chem. **234**, 148 (1959b).
ALLEN, W. W.: *American Chemical Paint Company* Tech. Service Data Sheet H 60 (1955).
ALVERSON, R. A.: Annotated bibliography of aminotriazole. Farm and Home Division, *American Cyanamid Company*, N.Y. (1957).
ANDERSEN, O.: Studies on the absorption and translocation of amitrol by nut grass (*Cyperus rotundus* L.). Weeds **6**, 370 (1958).
ASHTON, F. M.: Fate of amitrole in soil. Weeds **11**, 167 (1963).
— Degradation of amitrole in soil. Proc. Ann. Meeting Weed Soc. Amer., St. Louis, Mo., p. 48 (1961).
BEILSTEIN: Handbuch der organischen Chemie. 4. Aufl. Bd. 26, S. 137. Berlin: Springer 1931.
BLACK, C. C., and E. G. RODGERS: Response of Pensacola bahiagrass to herbicides. Weeds **8**, 71 (1960).
BRUNS, V. F., and J. H. DAWSON: Effects of DCB, DCB-xylene mixtures, amitrol, and sodium salt of dalapon in irrigation water on corn and rutabagas. Weeds **7**, 333 (1959).
BURSCHEL, P., and V. H. FREED: The decomposition of herbicides in soils. Weeds **7**, 157 (1959).
CARTER, M. C., and A. W. NAYLOR: The formation, in vivo, of a complex between 3-amino-1,2,4-triazole and a glycine-serine derivative. Plant Physiol. **34**, suppl. *vi* (1959).
— — Metabolism of labeled 3-amino-1,2,4-triazole in plants. Bot. Gaz **122**, 138 (1960).
— — Studies on an unknown metabolic product of 3-amino-1,2,4-triazole. Physiol. Plant **14**, 20 (1961).
CASTELFRANCO, P.: Inhibition of some metalloprotein enzymes by 3-amino-1,2,4-triazole. Biochim. Biophys. Acta **41**, 485 (1960).
—, and M. S. BROWN: A hypothesis of amitrol action based on its behavior toward free-radical generating systems. Weeds **11**, 116 (1963).
—, and A. OPPENHEIM: Reversal of amitrole-induced chlorosis by riboflavin. Abstr. 1961 Meeting Weed Soc. Amer. p. 43 (1961).
— —, and S. YAMAGUCHI: Riboflavin-mediated photodecomposition of 3-amino-1,2,4-triazole in relation to chlorosis. Weeds **11**, 111 (1963).
COUNTS, B.: Loss of geotropism and modification of root hair development produced by amitrole. Weeds **9**, 329 (1961).
CRAFTS, A. S.: Absorption and migration of synthetic auxins and homologous compounds. In: Handbuch der Pflanzenphysiologie. Bd. 14, S. 1044. Berlin-Göttingen-Heidelberg: Springer 1961.
— Translocation of herbicides. Hilgardia **26**, 287 (1956).
— Amino triazole (ATA). Adv. Pest Control Research **1**, 65 (1957).
— Further studies on comparative mobility of labeled herbicides. Plant. Physiol. **34**, 613 (1959).
— Improvement of growth regulator formulation, in plant growth regulation. Ames, Iowa: Iowa State University Press 1961.
— The chemistry and mode of action of herbicides. New York-London: Interscience 1961.
—, and S. YAMAGUCHI: Comparative tests on the uptake and distribution of labeled herbicides by *Zebrina pendula* and *tradescantia*. Hilgardia **27**, 421 (1958).
CROSS, C. E.: Chemical weed control in Massachusetts cranberry bogs. Proc. N.E. Weed Control Conf. **3**, 126 (1949).
DANA, M. N.: Weed control in Wisconsin cranberry marshes. Proc. N. Central Weed Control Conf. **12**, 50 (1956).
— Selective herbicides for cranberries. Proc. N. Central Weed Control Conf. **13**, 63 (1957).
— Effect of amitrol sprays on the growth and development of the cranberry. Weeds **7**, 277 (1959).

DAVIS, F. S., E. J. PETERS, D. L. KLINGMAN, H. D. KERR, and O. H. FLETCHALL: Chemical control of wild garlic. Weeds 10, 281 (1962).
DAY, B. E., L. S. JORDAN, and R. T. HENDRIXON: The decomposition of amitrole in California soils. Weeds 9, 443 (1961).
DEMORANVILLE, I. E., and C. E. CROSS: Newest cranberry weed killer, aminotriazole. Cranberries 21, 16 (1957 a).
— — The effects of aminotriazole on cranberry vines and their fruit. Cranberries 22, (1), 11 (1957 b).
— — The effects of aminotriazole on cranberry vines and their fruits. Cranberries 22 (11), 11 (1958).
DERSCHEID, L. A., K. E. WALLACE, and R. L. NASH: Leafy spurge control with cultivation, cropping and chemicals. Weeds 8, 15 (1960 a).
— — — Russian knapweed control with cultivation, cropping and chemicals. Weeds 8, 268 (1960 b).
—, R. L. NASH, and G. A. WICKS: Thistle control with cultivation, cropping and chemicals. Weeds 9, 90 (1961).
DOLL, C. C.: Chemical weed control in young grapes. Weeds 8, 368 (1960).
DURHAM, W. F.: Analytical methods for pesticides and antibiotics in food. Public Health Lab. Reports 19, 47 (1961).
DU SHANE, G.: Cranberry smash. Science 130, 1447 (1959).
EGGERT, R. G.: Unveröffentlichte Beobachtung (1959).
EISENBARTH, H. J.: Die Möglichkeiten der chemischen Unkrautbekämpfung im Weinbau. Weinberg und Keller 12, 93 (1965).
ENNIS, W. B. Jr.: Response of corn to 3-amino-1,2,4-triazole. Proc. S.W. Weed Control Conf. 9, 71 (1956).
ERCEGOVICH, C. D.: Analysis and persistence of 3-amino-1,2,4-triazole in plant tissue and soil. Dissertation Abstr. 18, 378 (1958).
—, and D. E. H. FREAR: The fate of 3-amino-1,2,4-triazole in soils. J. Agr. Food Chem. 12, 26 (1964).
FANG, S. C., M. GEORGE, and TE CHANG YU: Metabolism of 3-amino-1,2,4-triazole-5-C^{14} by rats. J. Agr. Food Chem. 12, 219 (1964).
FEINSTEIN, R. N., S. BERLINER, and F. O. GREEN: Mechanism of inhibition of catalase by 3-amino-1,2,4-triazole. Arch. Biochem. Biophys. 76, 32 (1958).
FLEMMING, A. S.: Statement released at news conference in Washington, D.C., Nov. 9, 1959.
FRANK, P. A., R. H. HODGSON, and R. D. COMES: Evaluation of herbicides applicated to soil for control of aquatic weeds in irrigation canals. Weeds 11, 124 (1963).
FRANS, R. E., E. F. LIND, and W. E. LOOMIS: Kinetics of growth inhibition by herbicides. Plant Physiol. 32, 301 (1957).
FREDRICK, J. F., and A. C. GENTILE: The stability constant of the manganese-chelate of 3-amino-1,2,4-triazole. Arch. Biochem. Biophys. 91, 178 (1960 a).
— — The formation of the glucose derivative of 3-amino-1,2,4-triazole under physiological conditions. Physiol. Plant 13, 761 (1960 b).
— — The structure of the D-glucose-addukt of 3-amino-1,2,4-triazole: Infrared absorption studies. Arch. Biochem. Biophys. 92, 356 (1961).
FREED, V. H., and M. MONTGOMERY: The effect of surfactants on foliar absorption of 3-amino-1,2,4-triazole. Weeds 6, 386 (1958).
GOWING, D. P.: A method of comparing herbicides and assessing herbicide mixtures at the screening level. Weeds 7, 66 (1959).
GREEN, F. O., and R. N. FEINSTEIN: Quantitative estimation of 3-amino-1,2,4-triazole. Anal. Chem. 29, 1658 (1957).
GUERIN-DUMARTRAIT, E.: Action du 3-amino-1,2,4-triazol sur la croissance, la teneur en pigments et les intensités photosynthetique et respiratoire de *Chlorella pyrenoidosa* chick. Compt. Rend. 250, 3044 (1960).
HALL, W. C., S. P. JOHNSON, and C. L. LEINWEBER: Chemical defoliant and regrowth inhibitor in cotton. Texas Agr. Expt. Station Bull., p. 759 (1963).
— — — Aminotriazole — a new abcission chemical and growth inhibitor. Texas Agr. Expt. Station Bull., p. 789 (1954).

HAUSER, E. W.: Effects of amitrole on purple nutsedge at different growth stages. Weeds **11**, 181 (1963).
—, and J. T. THOMPSON: A study of the absorption and translocation of several chemicals in Johnson grass, and an evaluation of their effectiveness for its control under field conditions. Weeds **7**, 20 (1959).
Hazleton Laboratories: Cranberry fingerprints. Instrument News *(Perkin Elmer Corp.)* **11**, 10 (1959).
HEIM, W. G., D. APPLEMAN, and H. T. PYFROM: Production of catalase changes in animals with 3-amino-1,2,4-triazole. Science **122**, 693 (1955).
— — — Effects of 3-amino-1,2,4-triazole on catalase and other compounds. Amer. J. Physiol. **186**, 19 (1956).
HERRETT, R. A., and W. P. BAGLEY: The metabolism and translocation of 3-amino-1,2,4-triazole by Canada thistle. J. Agr. Food Chem. **12**, 17 (1964).
—, and A. J. LINCK: The influence of 3-amino-1,2,4-triazole on the carbohydrate balance and respiration in Canada thistle *(Cirsium arvense)*. Plant Physiol. **32**, suppl. 6 (1957).
— — Quantitative determination of 3-amino-1,2,4-triazole. J. Agr. Food Chem. **9**, 466 (1961 a).
— — Penetration of 3-amino-1,2,4-triazole in Canada thistle and field bindweed. Weeds **9**, 224 (1961 b).
HILL, E. R., W. H. LACHMAN, and D. N. MAYNARD: Translocation of amitrole in yellow nutsedge and its effect on seed germination. Weeds **11**, 160 and 165 (1963).
HILTON, J. L.: Effect of histidine on the inhibitory action of 3-amino-1,2,4-triazole. Weeds **8**, 392 (1960).
— Riboflavin nullification of inhibitory actions of 3-amino-1,2,4-triazole on seedling growth. Plant Physiol. **37**, 238 (1962).
HOFSTEE, B. M. J.: On the mechanism of inhibition of xanthine oxydase by the substrate xanthine. J. Biol. Chem. **216**, 235 (1965).
HOLLINGSWORTH, E. B., and W. B. ENNIS: Studies on nut grass response to 3-amino-1,2,4-triazole and cultural practices. Proc. S. Weed Control Conf. **9**, 204 (1956).
HOSHINO, M.: Effect of 3-amino-1,2,4-triazole on the experimental production of liver cancer. Nature **186**, 174 (1960).
INGRAHAM, L. L.: Biochemical mechanisms, p. 108. New York-London: Wiley 1962.
JACKSON, W. T.: Effect of 3-amino-1,2,4-triazole and l-histidine on rate of elongation of root hairs of *Agrostis alba*. Weeds **9**, 437 (1961).
JANSEN, L. L., W. A. GENTNER, and W. C. SHAW: Effects of surfactants on the herbicidal activity of several herbicides in aqueous spray systems. Weeds **9**, 381 (1961).
JORDAN, L.: Factors affecting the decomposition of amitrol in soils. Proc. Ann. Calif. Weed Control Conf., p. 78 (1960).
JUKES, T. H., and C. B. SHAFFER: Antithyroid effects of aminotriazole. Science **132**, 296 (1960).
KADISCH, E.: Neue Erfahrungen mit der mechanischen und chemischen Unkrautbekämpfung im Weinbau In: Deutsches Weinbau-Jahrbuch, S. 82 (1965).
KRÖLLER, E.: Untersuchungen zur Restmengen-Bestimmung von 3-Amino-1,2,4-triazol in Preiselbeeren. Dtsch. Lebensmittel Rdsch. **57**, 107 (1961).
LEE, O. C.: Two years results with aminotriazole on Canada thistles. Research Rept. N. Calif. Weed Control Conf. **12**, 5 (1955).
LEEFE, J. S., and R. P. LONGLEY: The control of weeds around young apple trees. Weeds **8**, 422 (1960).
LEONARD, O. A., and L. A. LIDER: Toxicity and translocation of herbicides supplied to grape rootings through solution culture. Amer. J. Enol. Viticulture **12**, 37 (1961).
LEVI, E.: Chemical control of *Typha angustifolia* L. var. *brownii*. Weeds **8**, 128 (1960).
MACRAE, I. C., and M. ALEXANDER: Microbial degradation of selected herbicides in soil. J. Agr. Food Chem. **13**, 72 (1965).
MARGOLIASH, E., and A. NOVOGRODSKY: A study of the inhibition of catalase by 3-amino-1,2,4-triazole. Biochem. J. **68**, 468 (1958).

MARGOLIASH, E., A. NOVOGRODSKY, and A. SCHEJTER: Irreversible reaction of 3-amino-1,2,4-triazole and related inhibitors. Biochem. J. **74**, 339 (1960).
MASSINI, P.: Uptake and translocation of 3-amino-1,2,4-triazole and 3-hydroxy-1,2,4-triazole in plants. Acta Botan. Neerl. **7**, 524 (1958).
— Synthesis of 3-amino-1,2,4-triazolylalanine from 3-amino-1,2,4-triazole in plants. Biochem. Biophys. Acta **36**, 548 (1959). Vgl. hierzu auch Acta Botan. Neerl. **12**, 64 (1963).
MCWHORTER, C. G.: Some effects of 3-amino-1,2,4-triazole on the respiratory activities of *Zea mays* L. Dissertation, Louisiana State Univ. 1958.
—, and W. K. PORTER, JR.: The effects of 3-amino-1,2,4-triazole on respiration and oxidative enzymes of corn leaves. Abstr. 1958 Meeting Weed Soc. Amer., p. 40 (1958).
— — Some effects of amitrol on the respiratory activities of *Zea mays*. Weeds **8**, 29 (1960).
MEGGITI, W. F., and R. J. ALDRICH: Amitrol for control of redwood in cranberries. Weeds **7**, 271 (1959).
MILLER, C. S., and W. C. HALL: Effects of aminotriazole salts and derivatives on cotton defoliation growth inhibition and respiration. Weeds **5**, 218 (1957 a).
— — Aminotriazole salts and aminotriazole-5-C^{14} studies on pigment production in cotton. Weeds **5**, 304 (1957 b).
— — Absorption and metabolism of aminotriazole in cotton. J. Agr. Food. Chem. **9**, 210 (1961).
MITCHELL, L. C.: Paper chromatography of 3-amino-1,2,4-triazole. J. Assoc. Official Agr. Chemists **43**, 87 (1960).
NAYLOR, A. W.: Complexes of 3-amino-1,2,4-triazole in plant metabolism. J. Agr. Food Chem. **12**, 21 (1964).
NYLUND, R. E.: Pre- and postemergence weed control in garden beets. Proc. N. Central Weed Control Conf. **11**, 105 (1954).
ONLEY, J. H., R. W. STORHERR, and A. J. CULVER, JR.: An Oregon cranberry bog study for 3-amino-1,2,4-triazole residues in 1961—1962. J. Assoc. Official Agr. Chemists **46**, 996 (1963).
OSER, B.: The experimental induction of cancer by pesticide residues and food additives, its rationale and interpretation. Residue Reviews **1**, 6 (1962).
PALMER, R. D., and W. K. PORTER: The metabolism of nut grass (*Cyperus rotundus* L.). IV. The activities of certain enzymes from tubers treated with amitrol. Weeds **7**, 511 (1959).
PAVLYCHENKO, T. K.: Amino-triazole in controlling perennial weeds. Proc. N. Central Weed Control Conf. **12**, 37 (1955).
POTTS, K. T.: The chemistry of 1,2,4-triazoles. Chem. Reviews **61**, 87 (1961).
PYFROM, H. T., D. APPLEMAN, and W. G. HEIM: Catalase and chlorophyll depression by 3-amino-1,2,4-triazole. Plant Physiol. **32**, 674 (1957).
RACUSEN, D.: The metabolism and translocation of 3-aminotriazole in plants. Arch. Biochem. Biophys. **74**, 106 (1958).
RADWAN, M. A., C. R. STOCKING, and H. B. CURRIER: Histoauto-radiographic studies of herbicidal translocation. Weeds **8**, 657 (1960).
RALEIGH, S. M.: Quackgrass control. Proc. Meeting N. E. Weed Control Conf., p. 277 (Jan. 1955).
RIEPMA, P.: Preliminary observations on the breakdown of 3-amino-1,2,4-triazole in soil. Weed Research **2**, 41 (1962).
ROGERS, B. J.: Chlorosis in corn as induced by the herbicide 3-amino-1,2,4-triazole. Plant Physiol. **32**, suppl. vi (1957 a).
— Translocation and fate of aminotriazole in plants. Weeds **5**, 5 (1957 b).
— Action of 3-amino-1,2,4-triazole in plants. Hormolog *(Amchem Products Company)* 1957 c.
SCHWEIZER, E. E., and B. J. ROGERS: Effects of amitrole on acidsoluble nucleotides and ribonucleic acid in corn tissue. Weeds **12**, 310 (1964 a).
— — Structural requirements of amitrole for physiological activity. Weeds **12**, 7 (1964 b).
SEGAL, H. S.: The microdiffusion separation and determination of microgram quantities of thiocyanite in corn. J. Agr. Food Chem. **10**, 10 (1962).

SHAFFER, C. B.: Aminotriazole — acute and subacute toxicity. *American Cyanamid Company,* Central Medical Dept. 1956.
— Toxicology studies on aminotriazole. Proc. 1960 Ann. Calif. Weed Conf., p. 4 (1960).
SMITH, C. G.: Biosynthesis of riboflavin in *Eremothecium ashbyii.* Doctoral dissertation, Univ. of Chicago 1954.
STORHERR, R. W., and J. BURKE: Determination of 3-amino-1,2,4-triazole in crops. J. Assoc. Official Agr. Chemists **44,** 196 (1961).
—, and J. H. ONLEY: Cleanup and separation of 3-amino-1,2,4-triazole and metabolits from vegetable crops. J. Assoc. Official Agr. Chemists **45,** 382 (1962)
SUND, K. A.: Residual activity of 3-amino-1,2,4-triazole in soils. J. Agr. Food Chem. **4,** 57 (1956).
—, F. M. GORDON, and B. L. WALWORTH: Agricultural applications and soil residue studies of 3-amino-1,2,4-triazole. *American Cyanamid Company,* 18 pp. mimeo. 1955.
—, and H. N. LITTLE: Effect of 3-amino-1,2,4-triazole on the synthesis of riboflavin. Science **132,** 622 (1960).
—, E. C. PUTALA, and H. N. LITTLE: Phytotoxicity of herbicides. Reduction of 3-amino-1,2,4-triazole phytotoxicity in tomato plants. J. Agr. Food Chem. **8,** 210 (1960).
SUTHERLAND, G. L.: 3-Amino-*s*-triazole. In: G. ZWEIG, ed., Analytical methods for pesticides, plant growth regulators, and food additives. Vol. IV, p. 3. New York and London: Academic Press 1964.
TAFURO, A. J., and R. H. BEATTY: Progress report with greenhouse tests on Johnson grass seedlings using amizol as an additive with other herbicides. Ann. Meeting, S. Weed Control Conf., Jan. 1955.
TIMMONS, J. L., L. W. WELDON, and W. O. LEE: A study of factors which influence effectiveness of amitrol and dalapon on common cattail. Weeds **6,** 406 (1958).
UPCHURCH, R. P., G. R. LEDBETTER, and F. L. SELMAN: The interaction of phosphorus with the phytotoxicity of soil-applied herbicides. Weeds **11,** 36 (1963).
WEIR, R. J., O. E. PAYNTER, and J. R. ELSEA: Toxicology of 3-amino-1,2,4-triazole. Hormolog *(Amchem Products Company)* 1958.
WELCH, M. D., and C. VEATCH: Some effects of herbicides and season the rhizomes of quackgrass. Weeds **10,** 15 (1962).
WHITESIDE, J. S., and M. ALEXANDER: Measurement of microbiological effects of herbicides. Weeds **8,** 204 (1960).
WINKLER, W. O.: Determination of 3-amino-1,2,4-triazole in fresh cranberries and canned cranberry products. *U. S. Food and Drug Administration* "Bureau By-Lines" **1** (2), 49 (1959); modified by Seattle district, *U. S. Food and Drug Administration* under date of Nov. 9, 1959.
WOLF, F. T.: Influence of aminotriazole on the chloroplast pigments of wheat seedlings. Nature **188,** 164 (1960).
WORT, D. J., and B. C. LONGHMAN: The effect of 3-amino-1,2,4-triazole on the uptake, retention, distribution, and utilization of labelled phosphorus by young barley plants. Can. J. Bot. **39,** 339 (1961).
YAMAGUCHI, S., and A. S. CRAFTS: Comparative studies with labeled herbicides on woody plants. Hilgardia **29,** 171 (1959).
ZANARDI, D.: Eradication of secondary infections by *Peronospora tabacina* on tobacco plants through the use of desiccating agents. Tobacco **66,** 114 (1962).

Subject Index

GPSR Compliance

The European Union's (EU) General Product Safety Regulation (GPSR) is a set of rules that requires consumer products to be safe and our obligations to ensure this.

If you have any concerns about our products, you can contact us on ProductSafety@springernature.com

In case Publisher is established outside the EU, the EU authorized representative is:

Springer Nature Customer Service Center GmbH
Europaplatz 3
69115 Heidelberg, Germany

Batch number: 10844675

Printed by Printforce, the Netherlands